# Combustion and Pollution Control in Heating Systems

# V. I. Hanby

# Combustion and Pollution Control in Heating Systems

With 44 Figures

Springer-Verlag London Ltd.

Victor Ian Hanby, BSc, PhD, CEng, MInstE, MCIBSE
Department of Civil Engineering, University of Technology,
Loughborough, Leicestershire LE11 3TU, UK

ISBN 978-3-540-19849-9     ISBN 978-1-4471-2071-1 (eBook)
DOI 10.1007/978-1-4471-2071-1

British Library Cataloguing in Publication Data
Hanby, V. I.
 Combustion and Pollution Control in Heating Systems
 I. Title
 621.402

Library of Congress Cataloging-in-Publication Data
Hanby, V. I. (Victor Ian), 1942–
 Combustion and pollution control in heating systems / V.I. Hanby.
  p.   cm.
 Includes bibliographical references and index.


1. Heating—Equipment and supplies. 2. Hydrocarbons—Combustion.
3. Flue gases—Purification. I. Title.
TH7345.H36  1993                    93-27560
697—dc20                            CIP

© Springer-Verlag London 1994
Originally published by Springer-Verlag London Limited in 1994

Typeset by Richard Powell Editorial and Production Services, Basingstoke, Hants, RG22 4TX.
69/3830-543210 Printed on acid-free paper

# Preface

Combustion is very much an interdisciplinary topic, drawing together elements of chemistry, fluid mechanics and heat transfer. It is an ingredient in many undergraduate degree programmes, ranging from a pivotal role in fuel science through to a component part of courses in chemical, process and building services engineering.

For many students in those disciplines where combustion in heating plant is an important part of their studies, there are often problems in coming to grips with the basic principles underlying the combustion of hydrocarbon fuels. In particular, the concepts of chemical and related thermodynamic changes can prove difficult to assimilate.

The scientific literature dealing with combustion tends to be rather polarised, with a wealth of literature aimed at the specialist reader, but at a basic level the fundamentals of this important process are often treated rather tersely in textbooks on thermodynamics. The objective of this book is to provide an introduction to the basic principles of the combustion of hydrocarbon fuels in heating plant for buildings and industrial processes. In those chapters where practice in problem solving can make a positive contribution to understanding, some numerical problems have been included.

Acknowledging the ever-widening use of computers in technical education, a number of algorithms which can be easily coded up for solving numerical problems have been incorporated in the text. These can prove particularly useful in, for example, the calculation of certain fluid properties, either for use in hand calculation or for incorporation into larger programs.

It would be wholly inappropriate to leave out some discussion of the properties of hydrocarbon fuels and the equipment in which they are burned, but this has been deliberately confined to an outline treatment in Chapters 7, 8 and 9. There are many sources of information dealing with the practical aspects of combustion equipment, and reference is made in the text to a selection of appropriate books and guides to current practice.

Barrow-upon-Soar, 1994                                         V. I. Hanby

# Contents

**Chapter 1**

# The Combustion of Hydrocarbon Fuels

## 1.1 Introduction

It is difficult to overestimate the importance of the part played by combustion in the development of mankind. The key to this progress was the discovery of how to manage the combustion process; in other words, the ability to initiate combustion when required and to control and apply the resulting fire. There are three significant applications – the generation of power, the provision of heat for processes and the provision of heat for control of the built environment.

The earliest heating applications were directed towards the improvement of basic creature comforts: keeping habitable space warm and cooking food. The technology of the use of fire to control the thermal environment developed steadily and by Roman times the buoyancy forces generated by the hot combustion products were used to induce a flow of warmed air under the floors of buildings.

The development of combustion equipment for heating buildings followed a path of increasing efficiency and improving amenity: primitive 'plant' consisted of an open, central solid fuel fire with the flue consisting at best of a central opening in the roof. The health and safety aspects of this arrangement are difficult to imagine! There followed fireplaces with more effective flues formed into the building fabric, enclosed stoves, and finally the centralised boilers and heat distribution systems that are widely used today.

Combustion plant forms the heart of the great majority of heating systems in buildings (the exception being electrically heated installations) from flats through to district heating schemes. Such systems are fired by a wide variety of fuels, from natural gas through to municipal refuse, but common to all is the combustion of fuels containing the elements carbon and hydrogen – generically known as hydrocarbon fuels.

The treatment that follows deals with the fundamental aspects of the combustion process, as opposed to the technology of the plant which converts the chemical energy of the fuel into useful heat. Very broadly, these issues can be summarised as bringing together the air and fuel, consideration of the heat liberated by the process and dealing with the resulting products of combustion.

## 1.2  The Combustion Process

Combustion is a chemical reaction between a fuel and oxygen which is accompanied by the production of a considerable amount of heat (it is an *exothermic* reaction). Our everyday experience of flames can be used as a useful starting point for a discussion of the combustion reaction. The reaction has to be initiated by some source of high-temperature energy (ignition). We will not dwell on the significance of this at this stage except to note that the ease with which a fuel can be ignited is an important consideration in burner design.

The most obvious characteristic of the combustion process is the reaction zone, which is usually visible as a flame; the radiation emitted from the flame may be very intense, for example the characteristic yellow colour of an oil flame, or it may be quite weak as in the case of the flame from a gas hob on a cooker. At the completion of the combustion process we are left with the products of combustion.

In engineering terms, we can divide up the combustion of a fuel into several processes:

1. Bringing together the fuel and air (the reactants) in the correct proportions.
2. Igniting the reactants.
3. Ensuring that the flame burns in a stable manner and that combustion is complete.
4. Extracting useful heat from the process, and
5. Arranging for the safe disposal of the products of combustion.

By taking a very simple view of the way in which simple hydrocarbon fuels react with the oxygen present in the air a foundation can be laid for developing two of the most important basic procedures in the design and analysis of combustion systems, namely calculating the amount of air required for combustion and predicting the composition of the flue gases.

## 1.3  The Complete Combustion Reaction

We will simplify our view of the combustion reaction at this stage by assuming that all the fuel is burned to completion. Strictly speaking, this means that after a long period of time a chemical equilibrium has been reached, but in the present context it is sufficiently accurate to say that complete combustion has been achieved when no further reaction takes place – all the carbon in the fuel appears in the flue gases as carbon dioxide ($CO_2$) and all the hydrogen in the fuel is burned to water ($H_2O$).

As an illustration of the combustion reaction in the case of a simple but common gaseous hydrocarbon we can look at the combustion of methane ($CH_4$). Methane is the main component (about 95% by volume) of natural gas, and as implied by its chemical formula, a molecule of methane consists of a carbon atom bonded to four hydrogen atoms. The hydrogen atoms are equally spaced (three dimensionally) around the carbon atom, giving the molecule the characteristic tetrahedral structure shown in Fig. 1.1.

The bond structure of this molecule is quite stable, but during the combustion reaction it is broken (as is the bond structure of the reacting oxygen molecules); the carbon atom forms one molecule of carbon dioxide and the four hydrogen atoms form two molecules of water. This water is, of course, generated as a vapour which can subsequently condense to the liquid phase. Whenever we are dealing with combustion products the state of the water produced is important and it is essential to be quite clear about which phase is present.

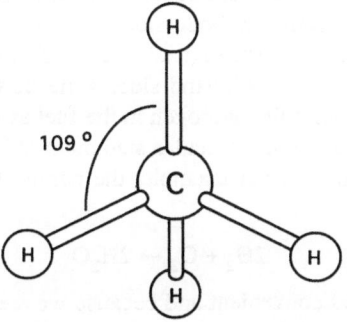

**Figure 1.1** The methane molecule

This bonding 'rearrangement' which takes place when the fuel and air react can be regarded as producing species (carbon dioxide and water vapour) which are thermo-dynamically at a lower energy level than the reactants. The transformation to a lower energy level is responsible for the exothermic nature of the reaction. A diagrammatic representation of the process is shown in Fig. 1.2 where it can be seen that it is necessary to get the reacting mixture over an energy 'hump' before the lower-energy products can be formed. This very simplified conceptual model will prove useful later on when considering the quantification of the amount of heat released when a fuel is

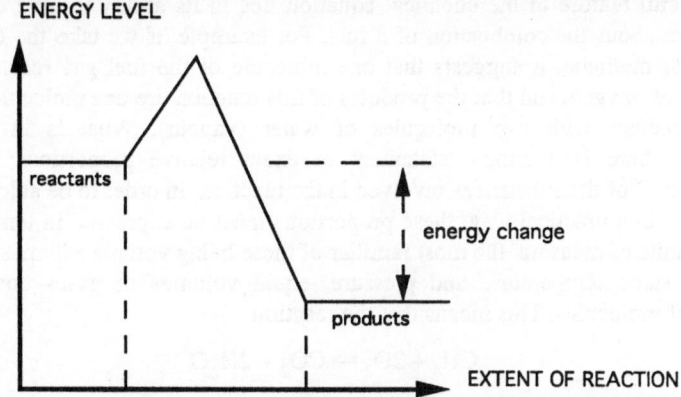

**Figure 1.2** Energy profile of the combustion reaction

burned.

The description of this process is, of course, succinctly described by writing down the reaction in the form of a simple chemical relationship:

$$CH_4 + 2O_2 \rightarrow CO_2 + 2H_2O$$

This relationship is often referred to as a chemical equation, with the arrow ( $\rightarrow$ ) sometimes replaced by an equals sign ( $=$ ). The arrow notation will be used here as, firstly, it shows graphically the expected direction of the chemical change and, secondly, with a small modification it enables us to indicate the existence of a chemical equilibrium condition.

The example reaction is, of course, balanced in the sense that equal numbers of atoms of each element appear on both sides of the relationship. Balancing the expression is easily done by inspection (trial and error); the usual starting point is to take all the carbon atoms on the left-hand side, write down the resulting number of molecules of $CO_2$, then balance the hydrogen in the fuel as water on the product side.

If we regard the equation as a molecular statement, all the numbers which prefix each species should be integers. For example, the combustion of hydrogen would be written:

$$2H_2 + O_2 \rightarrow 2H_2O$$

However, this is not always convenient and because we are in practice only interested in the relative proportions of the species present, it is perfectly acceptable to use fractional multipliers:

$$H_2 + \tfrac{1}{2}O_2 \rightarrow H_2O$$

The latter form is generally more convenient and will be used in this book.

# 1.4  Quantification of the Combustion Reaction

A very useful feature of the chemical equation lies in its ability to give quantitative information about the combustion of a fuel. For example, if we take the combustion reaction for methane, it suggests that one molecule of the fuel gas reacts with two molecules of oxygen, and that the products of this reaction are one molecule of carbon dioxide together with two molecules of water (vapour). What is in fact being represented here is a simple statement of exact relative proportions (known as *stoichiometry*) of the substances involved in the reaction. In order to be able to use this information in a practical way, these proportions must be expressed in terms of some practical units of measure, the most familiar of these being volume and mass.

At the same temperature and pressure, equal volumes of gases contain equal numbers of molecules. This means that the reaction

$$CH_4 + 2O_2 \rightarrow CO_2 + 2H_2O$$

also shows that one volume of methane requires just two volumes of oxygen to produce complete combustion. This is referred to as a *stoichiometric mixture* of the fuel and oxygen. Furthermore, if the products are at the same pressure as the reactants (which is normally the case in heating plant applications) and are brought to the same temperature, we will have one volume of carbon dioxide and either two volumes of water if it is in the vapour phase, or zero volume if it is liquid water, as solids and liquids are considered as having negligible specific volume compared to gases. The equation above can be quantified as:

$$CH_4 + 2O_2 \rightarrow CO_2 + 2H_2O$$

1 vol   2 vols      1 vol   2 (or 0) vols

The combustion reaction of a gaseous fuel is thus easily expressed in volumetric terms. We can also describe these quantitative relationships in mass terms by using the relative masses of the atoms present. More colloquially known as *atomic weights*, the values are expressed relative to the datum element oxygen at 16. For our purposes, it is sufficiently accurate to take the numerically rounded relative atomic masses of the elements which are important in hydrocarbon combustion, as given in Table 1.1.

**Table 1.1** Approximate relative atomic masses (atomic weights)

| Element | Symbol | Atomic weight |
|---------|--------|---------------|
| Carbon | C | 12 |
| Hydrogen | H | 1 |
| Oxygen | O | 16 |
| Nitrogen | N | 14 |
| Sulphur | S | 32 |

If the relative mass of each molecule in the combustion equation is expressed in terms of these atomic weights we have a basis for finding the relative masses of the substances involved. Taking as a basis each molecular weight expressed in kg:

$$CH_4 + 2O_2 \rightarrow CO_2 + 2H_2O$$

16 kg   64 kg      44 kg      36 kg

or

1 kg      4 kg      2.75 kg   2.25 kg

A third basis for quantification of the combustion reaction, and one which is perhaps the most generally applicable, is the use of molar quantities (this will perhaps seem the most natural to those readers who still recall some of their chemistry). One mole of any substance (in S.I. units) is its molecular (or atomic) weight expressed in grams (the term 'mole' is sometimes abbreviated to 'mol'). A mole of carbon (C) is therefore 12 g, one mole of oxygen ($O_2$) is 32 g and one mole of carbon dioxide ($CO_2$) 44 g. These are rather small quantities and it generally more convenient to use the molecular weight expressed in kilograms, the kilogram-mole, abbreviated as kg-mole, kmole or kmol.

The usefulness of molar quantities stems from the fact that 1 mole of any compound contains the same number of molecules ($6.0247 \times 10^{23}$ – the Avogadro number) hence proportions expressed in molar terms, e.g.

$$CH_4 + 2O_2 \rightarrow CO_2 + 2H_2O$$
$$\text{1 mole} \quad \text{2 moles} \quad \text{1 mole} \quad \text{2 moles}$$

are simple integer or rational quantities which are independent of the phase of the substance. This last point can be significant as the combustion of carbon

$$C + O_2 \rightarrow CO_2$$

cannot be meaningfully represented in volumetric terms as carbon is a solid fuel. Quantified on a mass basis it is apparent that 1 kg of carbon requires 2.66 (recurring) kg oxygen for complete combustion, which generates 3.66 (recurring) kg carbon dioxide. Expressed in moles, however, there is one mole of each reactant and one mole of product.

It is not really possible to generalise on which units should be chosen as a convenient basis for quantifying combustion processes. Because of its general applicability a molar basis would seem to be the most appropriate, but in practice volume and mass are both used extensively. It is probably fair to say that a reader with a predominantly scientific background would feel comfortable working in moles, whereas an engineer would be more familiar with mass and volume.

There is always a choice of the quantification basis, and this is usually based on expediency. For instance, the units of the available data may influence the choice, such as the specific heats of gases which may be available in kJ kg$^{-1}$ K$^{-1}$. The unit of supply for a gaseous fuel is normally volume, which may make a volumetric basis more convenient. In the treatment which follows, any of these three bases may be used in calculations and frequently more than one basis may be used in a procedure – the choice is simply that which is seen as most appropriate by the author.

# Chapter 2

# Stoichiometric Calculations

## 2.1 Applications of the Combustion Equation

There are two important practical applications for the treatment of the combustion equation developed in the previous chapter. Firstly, the ideal (stoichiometric) proportions in which the fuel combines with oxygen are used as a basis for finding the correct air supply rate for a fuel. Secondly, a knowledge of the expected composition of the combustion products is useful during the design, commissioning and routine maintenance of a boiler installation.

The flue gas composition is essential information if the flue system is to be correctly designed. Measurements of the flue gas composition during combustion are used to verify correct operation of the burner when the plant is being commissioned. Continuous monitoring of one or more of the flue gas constituents is often used to control the burner during normal operation and, finally, on-site measurements of flue gas composition and temperature are used as a basis for calculating the efficiency of the boiler at routine maintenance intervals.

## 2.2 Combustion Air Requirements: Gaseous Fuels

The supply of gaseous fuels is generally metered by volume and the chemical formulae of gaseous fuels are usually available. This means that in calculating the air required for combustion it is most convenient to work on a volumetric basis. As a starting point, let us consider as an example the stoichiometric air requirement for the fuel considered in the previous chapter – methane. The stoichiometric combustion reaction is

$$CH_4 + 2O_2 \rightarrow CO_2 + 2H_2O$$

which shows that each volume (normally 1 m$^3$) of methane requires 2 volumes of oxygen to complete its combustion. If the proportion of oxygen by volume in air is known it is then a simple matter to calculate the volume of air required. Air contains traces of many gases, and the concentrations of most of these change with time. Examples of species which change are carbon dioxide, hydrocarbons and oxides of nitrogen. If we ignore the components which are present in the parts per million range, air consists of about 0.9% by volume argon, 78.1% nitrogen and 20.9% oxygen (ignoring water vapour). Carbon dioxide is present at 0.034%. Argon is an inert gas and so for the purposes of combustion calculations the composition of air is approximated as a simple mixture of oxygen and nitrogen:

Oxygen     21%
Nitrogen   79%

We can now write the complete relationship for stoichiometric combustion:

$$CH_4 + 2O_2 + 7.52N_2 \rightarrow CO_2 + 2H_2O + 7.52N_2$$

(as the volume of nitrogen will be $2 \times 79 \div 21 = 7.52$). The nitrogen in the air is inert and passes through the flame into the combustion products. In fact a very small amount is oxidised but the resulting oxides of nitrogen (generally known as $NO_x$) are not formed in sufficient quantities to concern us here. Oxides of nitrogen are, of course, highly significant in terms of atmospheric pollution.

It can be seen straight away that the complete combustion of one volume of methane will require $(2 + 7.52 = 9.52)$ volumes of air, so the stoichiometric air-to-fuel (A:F) ratio for methane is 9.52. This ratio represents the starting point in combustion calculations, because in any practical situation more than the stoichiometric air quantity is supplied. This is because in practice it is impossible to obtain complete combustion under stoichiometric conditions. Incomplete combustion is a waste of energy and it leads to the formation of carbon monoxide, an extremely toxic gas, in the products.

*Excess air* is expressed as a percentage increase over the stoichiometric requirement, so if methane were to be burned with 20% excess air, the air supply would be 11.42 volumes of air per volume of fuel gas. It is easy to show that excess air is defined by:

$$\frac{(\text{actual A:F ratio} - \text{stoichiometric A:F ratio})}{\text{stoichiometric A:F ratio}} (\times 100\%)$$

The provision of excess air, however, should be limited to the minimum necessary to ensure good combustion of the fuel. The excess air supplied to the burner passes through the combustion zone as a 'passenger', and in doing so it leaves the system at a higher temperature than that at which it entered. Excess air will always reduce the efficiency of a combustion system. It is sometimes convenient to use the term *excess air ratio*, defined as

$$\frac{\text{actual A:F ratio}}{\text{stoichiometric A:F ratio}}$$

which is equal to $(1 + \text{excess air})$, expressed as a fraction.

Where sub-stoichiometric (fuel-rich) air-to-fuel ratios may be encountered, for instance, in the primary combustion zone of a low-$NO_x$ burner, the *equivalence ratio* is often quoted. This is given by:

$$\frac{\text{stoichiometric A:F ratio}}{\text{actual A:F ratio}}$$

# 2.3  Flue Gas Composition – Gaseous Fuels

The stoichiometric combustion equation for methane shows that the composition of the combustion products is

$$
\begin{array}{lll}
1 & \text{volume} & CO_2 \\
7.52 & \text{volumes} & N_2 \\
2 & \text{volumes} & H_2O
\end{array}
$$

giving a total product volume, per volume of fuel burned, of 10.52 if water is in the vapour phase, or 8.52 volumes if the water is condensed to a liquid. The two cases are usually abbreviated to 'wet' and 'dry' (with respect to the gas mixture). The proportion of carbon dioxide in this mixture is therefore

$$
\frac{1}{10.52} \times 100 = 9.51\% \text{ wet and } \frac{1}{8.52} \times 100 = 11.74\% \text{ dry}
$$

Not all the water vapour will condense out of the flue gases if they are cooled to around room temperature: the gas mixture will, in fact, be saturated with water vapour. In order to take this into account, it is necessary to find the partial pressure of the water vapour and to compare this with the saturation values obtained from steam tables. For this to be meaningful, it is necessary to know how much water vapour was present in the combustion air itself, but this level of detail is not relevant here.

It is perhaps appropriate to point out at this stage that the instruments used to measure the composition of flue gases remove water vapour from the mixture and hence give a dry reading, so the dry flue gas composition is usually of greater usefulness. Having described the basis of the technique, the calculation of the product composition is illustrated by considering the combustion of methane with 20% excess air.

The excess air $(0.2 \times 9.52)$ of 1.9 volumes will appear in the flue gases as $(0.21 \times 1.9) = 0.4$ volumes of oxygen and $(1.9 - 0.4) = 1.5$ volumes of nitrogen. The flue gas produced by the combustion of unit volume of methane with 20% excess air will contain the extra oxygen (0.4 volume) and a total of $(7.52 + 1.5) = 9.02$ volumes of nitrogen. The complete composition will be

| Constituent | Vol/Vol Methane |
|-------------|-----------------|
| $CO_2$ | 1 |
| $O_2$ | 0.4 |
| $N_2$ | 9.02 |
| $H_2O$ | 2 |

giving a total product volume of 12.42 (wet) or 10.42 (dry). The resulting composition of the flue gases, expressed as percentage by volume, is:

| Constituent | % Vol (dry) | % Vol (wet) |
|---|---|---|
| $CO_2$ | 9.6 | 8.1 |
| $O_2$ | 3.8 | 3.2 |
| $N_2$ | 86.6 | 72.6 |
| $H_2O$ | – | 16.1 |

## Example 2.1

A gas consists of 70% propane ($C_3H_8$) and 30% butane ($C_4H_{10}$) by volume. Find:

(a)  The stoichiometric air-to-fuel ratio and
(b)  The percentage excess air present if a dry analysis of the combustion products shows 9% $CO_2$ (assume complete combustion).

The combustion reactions for propane and butane are:

$$C_3H_8 + 5O_2 + 18.8N_2 \rightarrow 3CO_2 + 4H_2O + 18.8N_2$$

$$C_4H_{10} + 6.5O_2 + 24.5N_2 \rightarrow 4CO_2 + 5H_2O + 24.5N_2$$

It is often better to include the nitrogen on both sides if the problem involves consideration of the combustion products.

### Stoichiometric Air Requirement

Starting on the basis of 1 volume of the fuel gas, the propane content requires

$$0.7 \times (5 + 18.8) = 16.7 \text{ vols air}$$

and the butane requires

$$0.3 \times (6.5 + 24.5) = 6.3 \text{ vols air}$$

Hence the stoichiometric air-to-fuel ratio is *23:1*.

### Excess Air

The combustion products (dry) will contain

$$(0.7 \times 3) + (0.3 \times 4) = 3.3 \text{ vols } CO_2$$

$$(0.7 \times 18.8) + (0.3 \times 24.5) = 20.5 \text{ vols } N_2$$

plus $v$ volumes excess air, giving a total volume of products of $(23.8 + v)$. Given that the measured $CO_2$ in the products is 9%, we can write

$$\frac{9}{100} = \frac{3.3}{(23.8 + v)}$$

hence

$$v = 12.87 \text{ vols}$$

The stoichiometric air requirement is 23 vols so the percentage excess air is:

$$\frac{12.87}{23} \times 100 = 55.9\%$$

This part of the calculation shows that the excess air and actual air-to-fuel ratio can be worked out from the volumetric concentration of just one of the combustion products, knowing the composition of the fuel and also that combustion is complete (i.e. negligible carbon monoxide).

Many flue gas analysers measure oxygen concentration, both in order to control the excess air (oxygen trim control) and also for use in checking the performance of boilers and measuring their efficiency. Given the assumption of complete combustion, the percentage of oxygen in the flue gas can be used to evaluate the air-to-fuel ratio.

## 2.4    Combustion Air Requirements – Solid and Liquid Fuels

The principle of carrying out combustion calculations on solid and liquid fuels differs from that for a gaseous fuel because the fuels encountered in practice (coals and oils) have highly complex chemical compositions. The way in which the combustion equation is used reflects the available information on the analysis of the fuels. This takes the form of an element-by-element analysis (referred to as an *ultimate* analysis) which gives the percentage by mass of each element present in the fuel. An example of an ultimate analysis of a liquid fuel might be:

| Component | % by Mass |
|-----------|-----------|
| Carbon    | 86        |
| Hydrogen  | 14        |

In this and other books the components are abbreviated from a text string to a symbolic notation, so carbon is referred to as 'C' and hydrogen as '$H_2$'. It should be remembered that in this context the symbolic notation is a form of shorthand only. Molecular hydrogen is not present in the fuel but it is conventional to consider hydrogen, oxygen and nitrogen as being present in their molecular form and to use atomic symbols for other elements.

As a result of the fuel analyses being available in terms of percentages by mass, the calculations are performed on a mass basis. Each constituent is considered separately via its own combustion equation. Using the above oil as an exemplar, for the carbon

$$C \quad + \quad O_2 \quad \rightarrow \quad CO_2$$
$$12 \text{ kg} \qquad 32 \text{ kg} \quad \rightarrow \qquad 44 \text{ kg}$$

or for 1 kg of fuel

$$0.86 \quad 0.86 \times \frac{32}{12} \rightarrow 0.86 \times \frac{42}{12} \qquad \text{(kg)}$$

So each kg of oil requires 2.29 kg oxygen for combustion of its carbon and produces 3.15 kg $CO_2$ as product. Similarly

$$H_2 \quad + \quad \tfrac{1}{2}O_2 \rightarrow H_2O$$
$$2 \text{ kg} \qquad 16 \text{ kg} \rightarrow 18 \text{ kg}$$

or per kg of fuel

$$0.14 \quad 0.14 \times \frac{16}{2} \rightarrow 0.14 \times \frac{18}{2} \qquad \text{(kg)}$$

In order to burn the hydrogen content of the oil 1.12 kg oxygen are needed and 1.26 kg water is formed.

The total oxygen requirement is thus (2.29 + 1.12) or 3.41 kg. To obtain the air supply from this figure it is necessary only to know the proportions of oxygen and nitrogen in air by mass, as opposed to the volume values in the previous section. The composition of air by mass is, of course, readily available, but it is instructive to work it out from first principles here because the outcome of the combustion calculation will be a product composition by mass, so it necessary to be able to convert a gas composition from a mass to a volume basis and vice versa.

A given quantity of air consists of 21% by volume of oxygen. This means that 21% of the molecules in this sample are oxygen molecules, each one of which has a molecular weight ($M$) of 32. Similarly the remaining 79% of molecules are nitrogen with a molecular weight of 28. We can simply transform to a mass basis thus:

| Component | Vol fraction (vf) | vf × M | Mass fraction |
|---|---|---|---|
| Oxygen | 0.21 | 6.72 | $\frac{6.72}{28.84} = 0.233$ |
| Nitrogen | 0.79 | 22.12 | $\frac{22.12}{28.84} = 0.767$ |
| | | 28.84 | |

Notice that the weighted total at the bottom of the third column represents the 'average molecular weight' of air.

We can now establish that the 3.41 kg oxygen, which is the stoichiometric requirement, will be associated with:

$$3.41 \times \frac{0.767}{0.233} = 11.23 \text{ kg nitrogen}$$

The stoichiometric air-to-fuel ratio is thus 3.41 + 11.23 = 14.6:1.

# 2.5  Combustion Products – Solid and Liquid Fuels

The stoichiometric combustion products from combustion of the oil are:

$CO_2$     3.15 kg
$H_2O$     1.26 kg
$N_2$      11.23 kg

The combustion products would normally be needed as a volume percentage, so the reverse operation to that which was performed for air above is required. This is simply done by expressing each product in terms of kmoles produced, as the mole fraction of each gaseous component in the products is also its volume fraction. Hence if we require a dry volume percentage of the above products the following tabular procedure is convenient:

| Component | Mass/kg fuel | kmoles/kg fuel | Mole fraction |
|-----------|-------------|----------------|---------------|
| $CO_2$ | 3.15 | $\dfrac{3.15}{44}$ = 0.0716 | 0.151 |
| $N_2$ | 11.23 | $\dfrac{11.23}{28}$ = 0.4011 | 0.849 |
| | | 0.4727 | |

The stoichiometric combustion products are thus 15.1% $CO_2$ and 84.9% $N_2$.

Next, a worked example combustion calculation for a coal is presented. There are one or two points which should preface this example. Solid fuels, and many liquid fuels, contain compounds of sulphur. For the purposes of stoichiometric calculations this is assumed to burn to sulphur dioxide:

$$S + O_2 \rightarrow SO_2$$

In reality a mixture of sulphur dioxide and sulphur trioxide ($SO_3$) is produced, but it is conventional to assume combustion to $SO_2$ when calculating air requirements. The presence of sulphur in a fuel will be discussed elsewhere, as it is a major factor in causing air pollution and also oxides of sulphur produce highly corrosive acidic liquids if condensation should occur anywhere in the system.

Solid fuels and some oils produce ash when they burn. The percentage of ash in the fuel is part of the ultimate analysis and, as far as we are concerned at the moment, ash is simply treated as a totally inert substance.

A final point: many solid fuels contain small amounts of oxygen and nitrogen. Although these are 'locked up' in the form of complex molecules, the oxygen present in the fuel is considered to be available for burning the carbon, hydrogen and sulphur present, and the nitrogen in the fuel is taken to appear as gaseous nitrogen in the combustion products. All these points are illustrated in the following example.

## Example 2.2:   Combustion Calculation for a Coal

A coal has the following ultimate analysis:

|  | % by mass |
|---|---|
| Carbon | 90 |
| Hydrogen | 3 |
| Oxygen | 2.5 |
| Nitrogen | 1 |
| Sulphur | 0.5 |
| Ash | 3 |

Calculate:

(a) the volumetric air supply rate required if 500 kg h$^{-1}$ of coal is to be burned at 20% excess air and

(b) the resulting % $CO_2$ (dry) by volume in the combustion products.

Lay out the calculation on a tabular basis using 1 kg coal:

| | Mass (per kg) | $O_2$ Reqd | | Products | |
|---|---|---|---|---|---|
| Carbon | 0.9 | $0.9 \times \dfrac{32}{12}$ | = 2.4 | $0.9 \times \dfrac{44}{12}$ | = 3.3 |
| Hydrogen | 0.03 | $0.03 \times \dfrac{16}{12}$ | = 0.24 | $0.03 \times \dfrac{18}{12}$ | = 0.27 |
| Sulphur | 0.005 | $0.005 \times \dfrac{32}{32}$ | = 0.005 | $0.005 \times \dfrac{64}{32}$ | = 0.01 |
| Oxygen | 0.025 | −0.025 | | − | |
| Nitrogen | 0.01 | − | | 0.01 | |
| Ash | 0.03 | − | | − | |

Oxygen required to burn 1 kg coal = 2.4 + 0.24 + 0.005 −0.025 = 2.62 kg.

Air required = $\dfrac{2.62}{0.233}$ = 11.25 kg.

Actual air supplied = 11.25 × 1.2 = 13.5 kg.

Assuming a density for air of 1.2 kg m$^{-3}$ , the flow rate will be:

$$13.5 \times \frac{500}{1.2 \times 3600} = 1.56 \, \text{m}^3 \, \text{s}^{-1}$$

To get the %$CO_2$ in the combustion products we need to know the amounts of oxygen and nitrogen in the flue gases.

Air supplied = 13.5 kg per kg coal, of which oxygen is

$13.5 \times 0.233 = 3.14$ kg, and nitrogen

$13.5 - 3.14 = 10.36$ kg.

The combustion products will thus contain:

$3.14 - 2.62 = 0.52$ kg $O_2$ and

$10.36 + 0.01 = 10.37$ kg $N_2$.

A second tabular procedure can now be used for the volumetric composition of the flue gases:

| Product | Mass/kg coal | Mol. wt. | kmoles/kg coal | % Volume |
|---------|--------------|----------|----------------|----------|
| $CO_2$  | 3.3          | 44       | 0.075          | 16.25    |
| $SO_2$  | 0.01         | 64       | 0.000156       | 0.03     |
| $O_2$   | 0.52         | 32       | 0.0162         | 3.51     |
| $N_2$   | 10.37        | 28       | 0.37           | 80.20    |
|         |              |          | 0.4614         |          |

Note that the number of kmoles of each constituent is given by its mass divided by its molecular weight and that the % by volume is the number of kmoles of each constituent divided by the total number of kmoles of product from one kg of coal, expressed as a percentage.

## 2.6 Practical Significance of the Flue Gas Composition

The composition of the flue gas is important in a number of contexts, such as designing the flue or calculating the efficiency of the combustion device. The most common use for the flue gas composition is as an indicator of the air-to-fuel ratio at which the fuel is being burnt.

It is comparatively easy to make on-site measurements of the dry volumetric concentration of either carbon dioxide or oxygen in the flue gases. Either of these measurements can be used to calculate the air-to-fuel ratio (or excess air) if the composition of the fuel is known and the combustion of the fuel is complete, i.e. there is no carbon monoxide in the flue gas.

If the combustion of the fuel takes place with exactly the stoichiometric air requirement, there will be no oxygen present in the flue gas and the percentage of carbon dioxide will be a maximum. If we supply increasing amounts of excess air, the percentage of carbon dioxide will fall and the percentage of oxygen will increase.

Confining our interest at the moment to this important relationship between the flue gas composition and the excess air, the volume percentage of oxygen or carbon dioxide in the flue gas will be influenced by the level of excess air and also by the carbon:hydrogen ratio present in the fuel.

It simplifies matters if we consider for the moment that hydrocarbon fuels consist only of carbon and hydrogen. If pure carbon is burnt, the only combustion product is carbon dioxide, so each molecule of oxygen in the combustion air becomes a molecule of carbon dioxide in the flue gas. This means that the stoichiometric combustion of carbon will produce 21% by volume $CO_2$.

As the carbon:hydrogen ratio of the fuel decreases (i.e. the fuel contains increasing amounts of hydrogen) the stoichiometric air-to-fuel ratio will increase. This is because 1 kg carbon requires

$$\frac{32}{12} = 2.67 \text{ kg}$$

of oxygen for complete combustion but 1 kg hydrogen requires 16 kg oxygen. The percentage $CO_2$ in the flue gases will fall as the carbon:hydrogen ratio in the fuel decreases as

1. less carbon dioxide will be produced per kilogram of fuel and
2. the increased air requirement means that the carbon dioxide produced will be diluted by the extra nitrogen in the flue gas.

This effect is illustrated Table 2.1. The carbon:hydrogen ratios in fuels lie between the limits of 75:25 (methane) to around 95:5 (high carbon coals).

**Table 2.1** Carbon dioxide concentration in flue gases

| C : H (by mass) | | Stoichiometric %$CO_2$ |
|---|---|---|
| 100 | 0 | 21.00 |
| 95 | 5 | 18.67 |
| 90 | 10 | 16.62 |
| 85 | 15 | 14.81 |
| 80 | 20 | 13.19 |
| 75 | 25 | 11.73 |
| 70 | 30 | 10.42 |
| 65 | 35 | 9.23 |

There is a unique relationship between the composition of the flue gas and the excess air for any given fuel. This is illustrated in Fig. 2.1 for a natural gas with the following composition:

| Constituent | % by Volume |
|---|---|
| $CH_4$ | 92.6 |
| $C_2H_6$ | 3.6 |
| $C_3H_8$ | 0.8 |
| $C_4H_{10}$ | 0.2 |
| $C_5H_{12}$ (and higher) | 0.1 |
| $CO_2$ | 0.1 |
| $N_2$ | 2.6 |

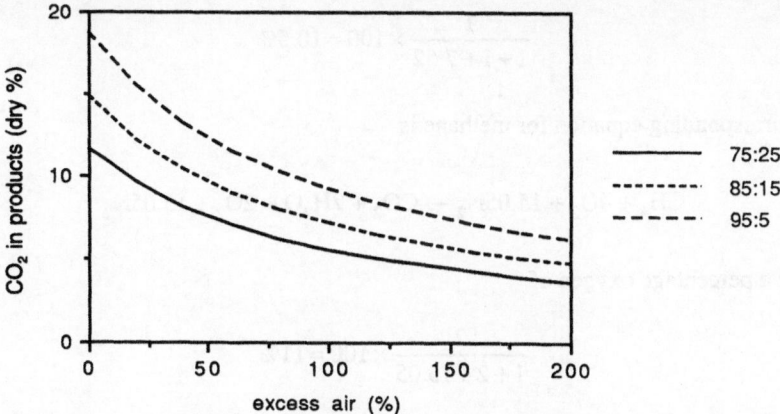

**Figure 2.1** Flue gas composition for natural gas

It is interesting to review the results of a series of such calculations for a range of hydrocarbon fuels. It would be expected that in every case the concentration of $CO_2$ would fall with increasing excess air, and that the curves for fuels with higher carbon:hydrogen ratios would lie above those for fuels with a lower value of this ratio.

Figure 2.2 shows a plot of the percentage $CO_2$ in the flue gases over a range of

**Figure 2.2** Carbon dioxide in combustion products

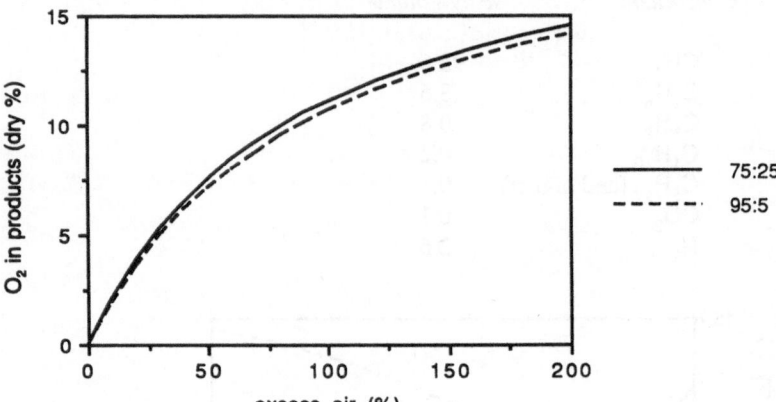

**Figure 2.3** Oxygen in combustion products

values of excess air for C:H ratios ranging from 75:25 to 95:5. The values decrease along a separate curve for each fuel, as would be expected. However, it can be seen in Fig. 2.3 that the relationship between the percentage oxygen in the flue gas and the excess air is very similar for a wide range of fuels.

It is to be expected that the values will converge at high excess air levels, but the fact that they are close at low values as well is due to the water vapour which is formed from the hydrogen in the fuel not being a component of the dry analysis.

As an example, take the combustion of two 'extreme' cases: one kmole of carbon and one kmole of methane. In each case we will consider 100% excess air. The combustion of carbon under these conditions is described by:

$$C + 2O_2 + 7.52N_2 \rightarrow CO_2 + O_2 + 7.52N_2$$

The percentage of oxygen in the flue gas is thus:

$$\frac{1}{1+1+7.52} \times 100 = 10.5\%$$

The corresponding equation for methane is

$$CH_4 + 4O_2 + 15.05N_2 \rightarrow CO_2 + 2H_2O + 2O_2 + 15.05N_2$$

giving a percentage oxygen of

$$\frac{2}{1+2+15.05} \times 100 = 11\%$$

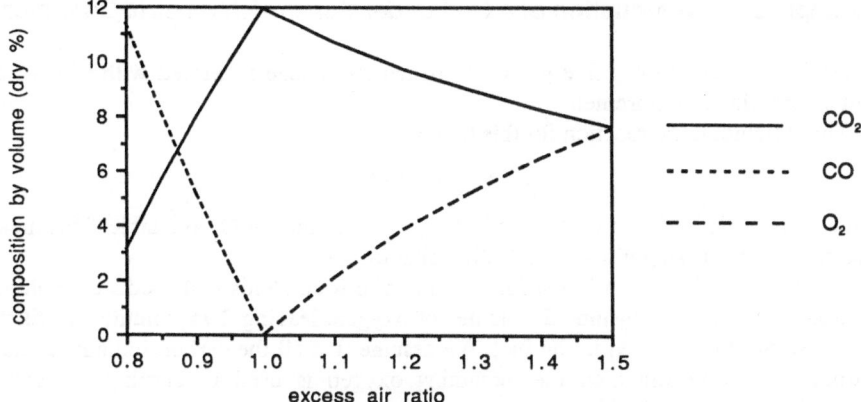

**Figure 2.4** Sub-stoichiometric combustion of natural gas

## 2.7  Sub-stoichiometric Combustion

Burning a fuel with less than the overall stoichiometric air requirement is not generally beneficial as some of the energy in the fuel is wasted and also there are serious implications as far as emissions in the flue gases are concerned. However, there are circumstances in which localised fuel-rich combustion can take place, such as where combustion of the fuel is a two-stage process with secondary air added downstream of the primary combustion zone.

Combustion of a fuel with less than the stoichiometric air requirement is a complex situation which involves chemical equilibria and reaction kinetics; however a simplified mechanism is conventionally assumed which enables a straightforward estimation of the products of combustion to be carried out. This mechanism consists of the following sequence of events:

1. The available oxygen firstly burns all the hydrogen in the fuel to water vapour.
2. All the carbon in the fuel is then burned to carbon monoxide.
3. The remaining oxygen is consumed by burning carbon monoxide to carbon dioxide.

The procedure outlined above is only applicable if there is enough oxygen to accomplish stages 1 and 2. Its implications are shown in the example of Fig. 2.4 where the combustion products for a typical natural gas are plotted as a function of the excess air ratio. It can be seen that as the air supply falls below the stoichiometric requirement (illustrated here by excess air ratios below unity) the percentage of carbon monoxide in the flue gas increases very quickly. The example given below illustrates the calculation procedure.

### Example 2.3: Combustion of a Fuel under Sub-stoichiometric Conditions

Estimate the wet and dry flue gas composition if propane is burned with 95% of the stoichiometric air requirement.

The stoichiometric reaction for this fuel is:

$$C_3H_8 + 5O_2 \rightarrow 3CO_2 + 4H_2O$$

On a volumetric basis we have $(5 \times 0.95) = 4.75$ volumes of $O_2$ available. This means that the accompanying nitrogen is 17.87 volumes.

Firstly all the hydrogen in the fuel is burned to water. This will produce 4 volumes of water vapour and consume 2 volumes of oxygen, leaving 2.75 volumes for further combustion of the carbon in the fuel. We assume that all the carbon initially burns to carbon monoxide and then the remaining oxygen is used in burning the carbon monoxide to carbon dioxide.

Burning the carbon to CO will produce 3 volumes of CO and use up 1.5 volumes of oxygen, leaving $(2.75 - 1.5) = 1.25$ volumes of oxygen for further combustion. This reaction is

$$CO + \tfrac{1}{2}O_2 \rightarrow CO_2$$

so 1.25 volumes oxygen can burn 2.5 volumes of carbon monoxide, producing 2.5 volumes of carbon dioxide. The remaining carbon monoxide is therefore $(3 - 2.5) = 0.5$ volume.

The products of combustion are thus:

| | |
|---|---|
| $N_2$ | 17.87 volumes |
| CO | 0.5 |
| $CO_2$ | 2.5 |
| $H_2O$ | 4.0 |
| | ——— |
| *Total* | 24.87 |

giving the percentage compositions:

| | Wet (%) | Dry (%) |
|---|---|---|
| $N_2$ | 71.9 | 85.6 |
| CO | 2.0 | 2.4 |
| $CO_2$ | 10.0 | 12.0 |
| $H_2O$ | 16.1 | – |

## 2.8 Problems

1. Propane is burned with 10% excess air. Calculate the volumetric air-to-fuel ratio and the dry composition of the flue gas.

[26.19:1; $CO_2$ 12.4%, $N_2$ 85.5%, $O_2$ 2.1%]

2. Calculate the stoichiometric air-to-fuel ratio for a gaseous fuel with the following volumetric composition:

   $H_2$      12.0%
   CO      29.0%
   $CH_4$     2.6%
   $N_2$      52.0%
   $CO_2$     4.0%

   [1.28:1]

3. A town gas has the following volumetric composition by volume:

   $H_2$      48%
   CO       5%
   $CH_4$     34%
   $CO_2$     13%

   What is the percentage of $CO_2$ in the flue gas (dry) when this fuel is burned with 25% excess air?

   [10%]

4. A distillate oil consists of 85% carbon (by mass) and 15% hydrogen. A dry volumetric analysis of the combustion products showed 1% oxygen and negligible carbon monoxide. Determine the percentage excess air and the air-to-fuel ratio.

   [4.8%, 15.7:1]

5. During the commissioning of an oil-fired boiler it was found that in order to obtain complete combustion of the fuel it was necessary to maintain 4% $O_2$ in the flue gas at the minimum firing rate and 2% $O_2$ at the maximum firing rate. Calculate the excess air percentage at both firing rates. The analysis of the oil was: carbon 85%, hydrogen 11%, sulphur 4%.

   [22.1%, 9.9%]

6. A sample of coal has the following ultimate analysis by mass:

   carbon 81%, hydrogen 5%, oxygen 5%, nitrogen 2%, sulphur 2%, ash 5%

   For stoichiometric combustion, calculate the air-to-fuel ratio and the dry volumetric composition of the flue gas.

   [10.89:1; $CO_2$ 18.4%, $N_2$ 81.5%, $SO_2$ 0.17%]

7.  A coal has the following ultimate analysis by mass:
    carbon 86%; hydrogen 5%; nitrogen 2%; oxygen 2%; ash 5%

    A dry volumetric analysis of the flue gas when the coal was burned gave a $CO_2$ reading of 15% and no measurable CO. Calculate the air-to-fuel ratio under these conditions.

    [14.1:1]

# Chapter 3

# Heat Release in Combustion

## 3.1 Introduction

We have until now been describing and quantifying the chemical changes which take place when a mixture of fuel and air burns. It is equally important to be able to quantify the thermal changes which take place on the combustion of a fuel – hence we

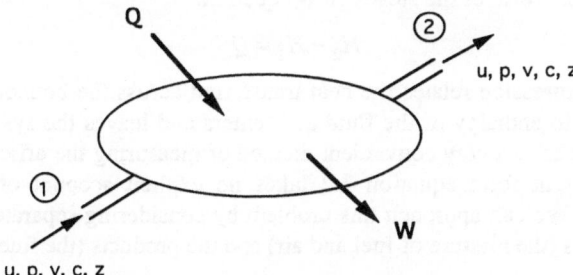

**Figure 3.1** Open system with combustion

must look at the thermodynamics of the combustion process.

We can envisage an elementary open system within a boundary containing a working fluid, which is initially a mixture of fuel and air (Fig. 3.1). There are two limiting cases which will cover the combustion of this mixture; the mixture could be ignited and burned at constant pressure (as occurs in boilers), or combustion could take place at constant volume. Because we are concerned here with combustion in boilers and other heating systems, we will focus our attention on the case of combustion at constant pressure.

## 3.2 Constant-pressure Combustion

A combustion system must obey the First Law of Thermodynamics, but the application of this law may at first sight appear difficult owing to the change in chemical composition of the 'working fluid' from a mixture of fuel and air to the resulting combustion products. The application of the First Law to the open system of Fig. 3.1 gives, for each kg of working fluid, the following equation:

$$(U_2 - U_1) + (P_2V_2 - P_1V_1) + g(Z_2 - Z_1) + \tfrac{1}{2}(C_2^2 - C_1^2) = Q - W \qquad (3.1)$$

The terms on the left-hand side are respectively the change in internal energy of the fluid, the 'flow work' involved in propelling the fluid through the system, the change in potential energy of the fluid and finally its change in kinetic energy. The work done by the system $(W)$, in the case of a boiler, is zero, so the right-hand side of the equation represents the amount of heat transferred into the system (adopting the convention that work done by the system and heat transferred into the system are both positive quantities).

The changes in potential and kinetic energy of the fluid can be considered negligible, so equation 3.1 simplifies to:

$$(U_2 - U_1) + (P_2V_2 - P_1V_1) = Q$$

The enthalpy of the fluid, $H$, is given by

$$H = U + PV$$

leading to a final form of the steady flow equation:

$$H_2 - H_1 = Q \qquad (3.2)$$

This simple expression relates the heat transferred across the boundary of the system to the change in enthalpy of the fluid as it enters and leaves the system. We will see later how it leads to a very convenient method of measuring the efficiency of a boiler, but in its present form equation 3.2 takes no explicit account of the combustion process itself. We can approach this problem by considering separately the enthalpies of the reactants (the mixture of fuel and air) and the products (the flue gases).

## 3.3  Enthalpy of a Mixture of Gases

For the purposes of this discussion it will be assumed that the combustible mixture contains a gaseous fuel. The reactants can be treated as a homogeneous entity if ideal gas behaviour is assumed, hence a change in the enthalpy of the reactants can be calculated by summing the enthalpy changes of each of the constituent gases. The change in enthalpy of a gas as a function of temperature is given by

$$\Delta H = c_p(\Delta t)$$

which would at first sight appear to give a straight line relationship between enthalpy change and temperature difference. However, the temperature changes which occur in combustion are considerable; nitrogen, for example, could enter the combustion system at ambient temperature (15°C) and be heated in the flame to around 2000°C. At the lower temperature nitrogen has a specific heat at constant pressure of 1.04 kJ kg$^{-1}$ K$^{-1}$, rising to 1.30 kJ kg$^{-1}$ K$^{-1}$ at flame temperature. All gases have values of $c_p$ which increase with temperature, hence a property diagram relating the enthalpy of the reactants $H_R$ to their temperature will be a curve similar to that in Fig. 3.2. Remembering that the enthalpy change of the mixture can be obtained by summing the changes of each of its constituents, an equation describing the curve is

$$\Delta H_R = \Sigma (m.c_p)_R (\Delta t)$$

where the summing operation is done for each of the reactants, i.e. fuel gas, oxygen and nitrogen. It should be noted at this stage that if this equation is used to evaluate enthalpy changes over a significant temperature interval, account must be taken of the change in specific heat with temperature. For most purposes, an averaged value taken over the temperature interval is adequate, but the variation of specific heat with temperature is not linear and for accurate work tabulated enthalpy values should be used.

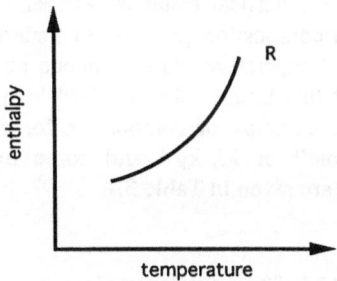

**Figure 3.2** Enthalpy of fuel/air mixture

# 3.4  Enthalpy of Combustion

Clearly the argument of the preceding section applies equally to the mixture of gases produced by the combustion process and so the enthalpy–temperature diagram for the products will be a curve of similar shape, described in this instance by:

$$\Delta H_P = \Sigma (m.c_p)_P (\Delta t)$$

These two curves, for reactants and products, can be related to one another by considering a fixed quantity of fuel and air at a given temperature. If this mixture is ignited and the resulting products returned to the initial temperature, clearly a quantity of heat will have left the system. The implication of this in terms of the First Law (equation 3.2) is that the enthalpy of the products is less than the enthalpy of the reactants at the same temperature. We can thus sketch two curves on an enthalpy–temperature diagram, one for the reactants and one for the products, with the curve for the products lying below that for the reactants (Fig. 3.3).

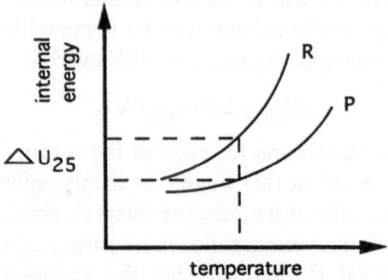

**Figure 3.3** Enthalpy of reactants and products

At any given temperature, the vertical distance between the two curves represents the enthalpy released by the combustion process. In general the magnitude of this quantity will depend on the temperature chosen, hence an arbitrary temperature is specified for the reporting of this figure. The standard temperature adopted is 25°C and the quantity $\Delta H_{25}$ is the enthalpy of combustion for the fuel in question. It is normally quoted in kJ kmole$^{-1}$ or kJ kg$^{-1}$ and some examples of enthalpy of combustion for gaseous fuels are given in Table 3.1.

**Table 3.1**  Standard enthalpies of combustion

| Fuel | $\Delta H_{25}$ (kJ kmole$^{-1}$) |
|------|------|
| Hydrogen (H$_2$) | −241 800 |
| Methane (CH$_4$) | −802 300 |
| Propane (C$_3$H$_8$) | −2 045 400 |
| Acetylene (C$_2$H$_2$) | −1 256 400 |

The combustion of all the above fuels will produce water in the flue gases, which can be considered as existing in either the liquid or vapour phases. The figure quoted for the enthalpy of combustion will clearly depend on the phase of the water in the combustion products. All the figures given above are for water in the vapour phase. Should the enthalpy of combustion be required with water in the liquid phase, then the latent heat of evaporation for the water vapour produced in the combustion products ($H_{fg}$) must be accounted for:

$$(\Delta H_{25})_f = (\Delta H_{25})_g - n \times H_{fg}$$

where $n$ represents the number of kmoles of water produced per kmole of fuel. The latent heat at 25°C has a value of 44 000 kJ kmole$^{-1}$ of water produced. The numerical difference between the two cases is significant, as the following example will show.

## Example 3.1

The value of $\Delta H_{25}$ for methane is $-802\,300$ kJ kmole$^{-1}$ with water in the vapour phase. Calculate $\Delta H_{25}$ for the case when all the water vapour is condensed.

The relevant stoichiometric combustion equation for methane is

$$CH_4 + 2O_2 \rightarrow CO_2 + 2H_2O$$

i.e. 2 kmoles of water vapour are produced for each kmole of fuel burnt. Hence the value for $\Delta H_{25}$ with water in the liquid phase is given by

$$(\Delta H_{25})_f = (\Delta H_{25})_g - 2 \times H_{fg}$$

$$= -802\,300 - 2 \times 44\,000$$

$$= -890\,300 \text{ kJ kmole}^{-1}$$

# 3.5  Constant-volume Combustion

The combustion of a fuel at constant volume, as a process, is of little practical significance in terms of the operation of heating plant. However, the heat released when a solid or liquid fuel is burnt is measured by burning a sample of the fuel at constant volume, hence it is important to look at the thermodynamics of this process as well as that of combustion at constant pressure.

# 3.6  Internal Energy of Combustion

The combustion of a mixture of fuel and air at constant volume is most easily envisaged as taking place inside a rigid closed container. The First Law energy equation for a closed system is:

$$(U_2 - U_1) = Q - W \qquad (3.3)$$

As the system boundary is fixed, no work can cross it, hence equation 3.3 becomes:

$$(U_2 - U_1) = Q \qquad (3.4)$$

The change in internal energy can be related to temperature rise by:

$$\Delta U = c_v(\Delta t)$$

Thus we can plot an internal energy–temperature diagram for the reactants and products of combustion in the same way as an enthalpy–temperature plot was made in Section 3.3. The curves of Fig. 3.4 are described by:

$$\Delta U_R = \Sigma(m.c_v)_R(\Delta t)$$

and

$$\Delta U_P = \Sigma(m.c_v)_P(\Delta t)$$

Once again, at any given temperature the internal energy of the products must be less than that of the reactants (since heat has left the system) and the difference between the two curves at the reference temperature of 25°C is referred to as the *internal energy of combustion* for that fuel.

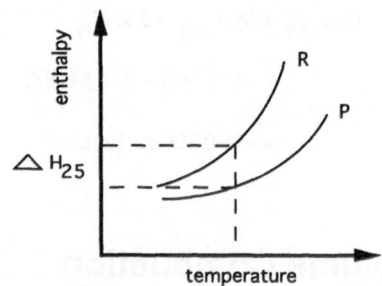

**Figure 3.4** Internal energy of reactants and products

A clear distinction has again to be made between the two phases of the water produced by burning the fuel. Each fuel will have two values for internal energy of combustion: one for water in the vapour phase and one for water in the liquid phase. The two values will differ, as in the case of constant-pressure combustion, because of the water produced by the combustion of the fuel, but for calculating the difference for constant-volume combustion the internal energy of evaporation ($U_{fg}$) of water at 25°C should be used. This value can be taken as $37\,602$ kJ kmole$^{-1}$.

## 3.7  Relationship between $\Delta H_{25}$ and $\Delta U_{25}$

It should be clear that the enthalpy of combustion is related to combustion at constant pressure and that the internal energy of combustion of a fuel is correspondingly related to a constant-volume process. These two quantities are also associated with fuel types, because calorific measurements are usually performed at constant pressure for a gaseous fuel, and at constant volume for liquid and solid fuels. As all types of fuels are burned at constant pressure in heating plants, the question naturally arises as to the nature and magnitude of the difference between the two values.

At the standard reference temperature of 25°C we can write

$$\Delta H_{25} = (H_P - H_R)_{25}$$

Noting that

$$H = U + PV$$

we get

$$\Delta H_{25} = (U_P - U_R) + \{(PV)_P - (PV)_R\}$$
$$= \Delta U_{25} + \{(PV)_P - (PV)_R\}$$

The $(PV)$ terms for solids and liquids are very small, so for any reactant or product in the gas phase

$$PV = nRT$$

hence

$$\Delta H_{25} = \Delta U_{25} + RT(n_P - n_R) \qquad (3.5)$$

Here $n_P$ represents the total number of kmoles of combustion products in the gas phase and $n_R$ represents the number of kmoles of gaseous reactants – clearly if $n_P$ and $n_R$ are equal (implying no volume change during combustion) then $\Delta H_{25}$ and $\Delta U_{25}$ will have the same value. In general there is a volume change accompanying the combustion of a fuel, but the difference between the two values is quite small, as shown by the following example.

## Example 3.2

$\Delta H_{25}$ for propane ($C_3H_8$) is –2 045 400 kJ kmole$^{-1}$ with water in the vapour phase. Calculate the corresponding internal energy of combustion.

We can write the stoichiometric equation for propane ignoring the presence of nitrogen and any excess oxygen as they will appear in both the reactants and the products:

$$C_3H_8 + 5O_2 \rightarrow 3CO_2 + 4H_2O$$

i.e. 6 kmoles of gaseous reactants become 7 kmoles of products so

$$n_P - n_R = 1$$

At the reference temperature of 298 K we get (from equation 3.5):

$$-2\,045\,400 = \Delta U_{25} + (1 \times 8.314 \times 298)$$
$$= \Delta U_{25} + 2478$$
$$\Delta U_{25} = -2\,047\,878\,\text{kJ kmole}^{-1}$$

In conclusion, the two numerical examples in this chapter illustrate that there is normally little numerical difference between the enthalpy and internal energy of combustion of a fuel. However, the phase of any water produced by combustion (liquid or vapour) is highly significant and it must be clearly associated with any quoted values.

# 3.8 Calorific Values

The terms 'enthalpy of combustion' and 'internal energy of combustion' have precise thermodynamic definitions and their use is appropriate where a high degree of scientific precision is required. However, in more practical situations engineers generally use calorific values as a measure of the heat released when unit quantity of a fuel is burned.

The calorific value of a fuel is an experimentally determined figure, namely the amount of heat released when a known quantity of fuel is burned under specified conditions. Experimental determinations of calorific value are made at both constant pressure and constant volume, so there is often some confusion at first acquaintance as to why, for example, the calorific value for a liquid fuel measured at constant volume differs from its internal energy of combustion. Such discrepancies are small and they arise from the actual conditions under which the experimental measurements are made; this will be apparent when the operation of the calorimeters used are described.

The calorific value of a fuel is normally expressed as kJ (or MJ) per $m^3$ (gaseous fuels) or kJ (MJ) per kg which is applicable to all types of fuel. The phase of the water produced is clearly important – if the calorific value includes the latent heat of condensation of the water produced it is usually referred to as the *gross calorific value*, whereas if the water is in the vapour phase, the term *net calorific value* is used. The expressions *higher calorific value* and *lower calorific value* are also in use, but are not so common.

A fuel can have four calorific values:

1. gross calorific value at constant volume;
2. gross calorific value at constant pressure;
3. net calorific value at constant volume;
4. net calorific value at constant pressure.

The numerical values of 1 and 3 correspond closely to the two values of $\Delta U_{25}$ while the values of 2 and 4 are close to the two values for $\Delta H_{25}$. The difference between the heat released when a fuel is burned at constant volume as opposed to constant pressure is very small, but it is very important to distinguish clearly between gross and net calorific values.

Given one of either the gross or net calorific values of the fuel, it is a straightforward matter to calculate the other value if the mass of water produced from burning unit quantity of the fuel is known. The value for $h_{fg}$ for water at 25°C is 2442 kJ kg$^{-1}$.

## Example 3.3

A liquid fuel consists of 86% carbon and 14% hydrogen (by mass). Its gross calorific value is 43.5 MJ kg$^{-1}$. Calculate the net calorific value.

1 kg fuel contains 0.14 kg hydrogen. The stoichiometric combustion equation for the hydrogen is

$$H_2 + \tfrac{1}{2}O_2 \rightarrow H_2O$$

hence 2 kg hydrogen produce 18 kg water vapour. The mass of water vapour produced by the combustion of 1 kg fuel is thus

$$0.14 \times 9 = 1.26 \text{ kg}$$

The net calorific value of the fuel is then

$$43.5 - 1.26 \times 2.442 = 40.42 \text{ MJ kg}^{-1}$$

The calorific value of a fuel derives from the heat released when the carbon and hydrogen in the fuel are burned. On a per kilogram basis, hydrogen has a much higher calorific value than carbon. Carbon has a calorific value of 32.8 MJ kg$^{-1}$ whereas molecular hydrogen has a net calorific value of 120.9 MJ kg$^{-1}$.

The calorific value of a hydrocarbon fuel cannot be predicted with any degree of accuracy from these figures, but there is an implicit suggestion that the calorific value of a fuel may be roughly related to the relative quantities of carbon and hydrogen from which it is made up. The broad trend is illustrated in Fig. 3.5 where the net calorific values of a range of fuels is plotted against their carbon:hydrogen ratio. The range runs from anthracite and bituminous coal (the highest C:H ratio) to natural gas (the lowest).

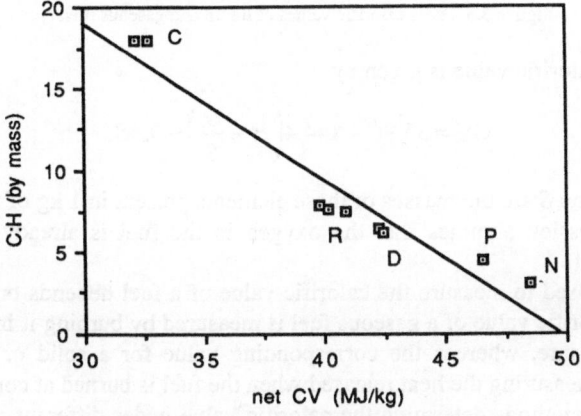

**Figure 3.5** Net calorific values of hydrocarbon fuels

A much more consistent trend is exhibited by the liquid and gaseous fuels, which have a much higher proportion of hydrogen than do coals. Figure 3.6 shows the gross calorific values of these fuels as a function of their C:H ratio. The points fall closely on a line, with the evident grouping of residual fuel oils (H, M and L), the distillate oil fuels (G and K) and finally the gaseous fuels propane (P) and natural gas (N).

The calorific values for solid fuels are considerably lower than the range for liquid and gaseous fuels (43–54 MJ kg$^{-1}$). Values for coals lie in the range 22–33 MJ kg$^{-1}$. This is partly due to their low hydrogen content (around 3–4% by mass) but it is also due to the presence of ash, oxygen and nitrogen in the fuel.

It is possible to obtain a rough estimate of the calorific value of a solid or liquid fuel from its ultimate analysis and the calorific values of its combustible constituents. For this purpose, the relevant gross calorific values are:

carbon     33.9 MJ kg$^{-1}$
hydrogen   144.4 MJ kg$^{-1}$
sulphur    9.4 MJ kg$^{-1}$

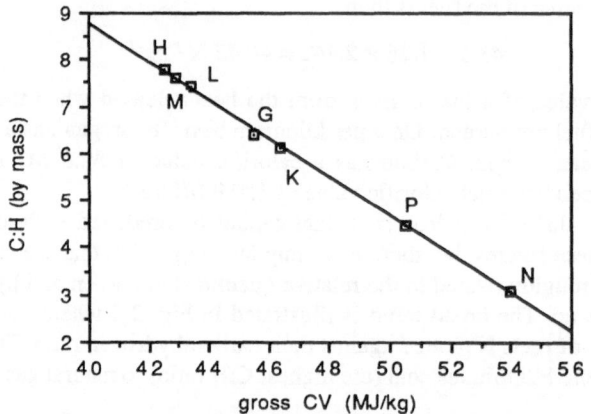

Figure 3.6 Gross calorific values of liquid and gaseous fuels

The estimated calorific value is given by

$$CV = 33.9C + 144.4\left(H - \frac{O}{8}\right) + 9.4S$$

where $C$, $H$, $O$ and $S$ are the masses of these elements present in 1 kg of fuel. Note that the above expression assumes that the oxygen in the fuel is already bound to the hydrogen present.

The method used to measure the calorific value of a fuel depends on the nature of the fuel: the calorific value of a gaseous fuel is measured by burning it in a calorimeter at constant pressure, whereas the corresponding value for a solid or liquid fuel is determined by measuring the heat released when the fuel is burned at constant volume. Although these methods determine the calorific value under different conditions, the figures obtained are generally applied regardless of the actual circumstances under which the fuel is burned, as the change in volume when most fuels are burned is small.

The calorific value of a gaseous fuel is measured in a device known as the Boys' calorimeter. The operation of this device is outlined diagrammatically in Fig. 3.7; it is basically a counterflow heat exchanger with the reacting fuel and air flowing in the opposite direction to the flow of cooling water. When steady-state conditions are obtained, the heat released by burning the fuel is calculated from the temperature rise of the cooling water and its mass flow rate. The apparatus is designed so that the average of the starting and finishing gas temperatures is close to the reference value of 25°C.

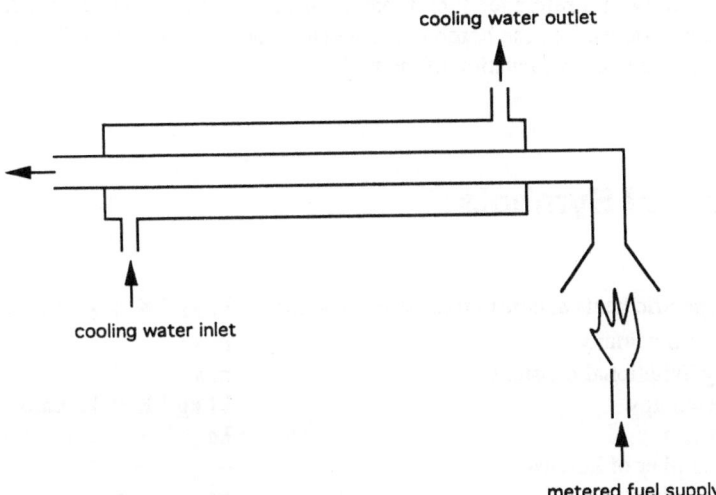

**Figure 3.7** Boys' calorimeter

This temperature is well below the dew point of the combustion products, so the value measured is the gross calorific value of the fuel. Not all the water vapour generated in the combustion of the fuel may condense out in the calorimeter, but this can be corrected for by supplying the calorimeter with near-saturated air.

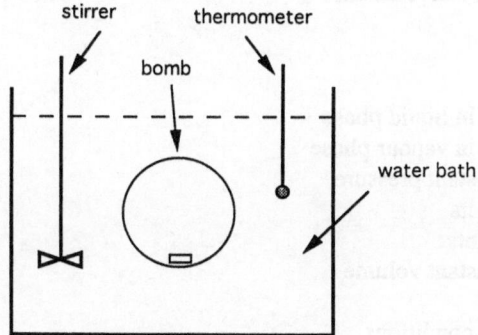

**Figure 3.8** Bomb calorimeter

The calorific value of a liquid or solid fuel is measured in the bomb calorimeter (Fig. 3.8). This device consists of a spherical vessel containing a small sample of the fuel in an atmosphere of pressurised oxygen. The bomb is immersed in an insulated water bath which acts as the calorimeter for the heat released when combustion is initiated. The temperature change in the water is measured by a sensitive thermometer, and knowing the thermal capacity of the apparatus the heat released can be determined.

Once again, the temperature rise is arranged to take place from just below, to just above the reference temperature of 25°C with a maximum rise of about 4°C. In practice, the thermal capacity of the apparatus is evaluated experimentally by burning a known quantity of a reference fuel, typically benzoic acid, in the device. A quantity of water is introduced into the bomb to ensure saturation, hence the condensation of all the water produced by combustion of the fuel.

## 3.9  List of Symbols

| | | |
|---|---|---|
| $c_p, c_v$ | specific heats at constant pressure, volume | kJ kg$^{-1}$ K$^{-1}$;  kJ kmole$^{-1}$ K$^{-1}$ |
| $C$ | fluid velocity | m s$^{-1}$ |
| $g$ | gravitational constant | m s$^{-2}$ |
| $H$ | enthalpy | kJ kg$^{-1}$ K$^{-1}$;  kJ kmole$^{-1}$ K$^{-1}$ |
| $m$ | mass | kg |
| $n$ | number of kmoles | – |
| $P$ | pressure | Pa |
| $Q$ | heat | kJ |
| $R$ | universal gas constant | kJ kmole$^{-1}$ K$^{-1}$ |
| $t$ | temperature | degrees Celsius |
| $T$ | temperature | degrees Kelvin |
| $U$ | internal energy | kJ kg$^{-1}$ K$^{-1}$;  kJ kmole$^{-1}$ K$^{-1}$ |
| $V$ | volume | m$^3$ |
| $W$ | work | kJ |
| $Z$ | height above datum | m |

*Subscripts*

| | |
|---|---|
| f | water in liquid phase |
| g | water in vapour phase |
| p | at constant pressure |
| P | products |
| R | reactants |
| v | at constant volume |
| | |
| 1 | initial conditions |
| 2 | final conditions |
| 25 | at standard reference temperature, 25°C |

## 3.10  Problems

1. An oil consists of 85% carbon and 15% hydrogen by mass. A bomb calorimeter test gave a gross calorific value of 46.90 MJ kg$^{-1}$. What would be the gross calorific value at constant pressure?

[46.99 MJ kg$^{-1}$]

2. A liquefied petroleum gas consists of 75% propane ($C_3H_8$) and 25% butane ($C_4H_{10}$) by volume. Calculate the net calorific value of this gas. The gross calorific values are 96.0 MJ m$^{-3}$ (propane) and 102.5 MJ m$^{-3}$ (butane).

[94.2 MJ m$^{-3}$]

3. Estimate the net calorific value of a coal consisting of 82% carbon, 6.5% hydrogen, 3.5% oxygen and 3% sulphur.

[35.4 MJ kg$^{-1}$]

# Chapter 4

# Flame Temperature

## 4.1 Energy Balance on a System

A simple steady-state thermal energy balance can be constructed around a constant-pressure combustion system (such as a boiler), which is useful in a number of contexts. Figure 4.1 shows the energy flows into and out of the system which are

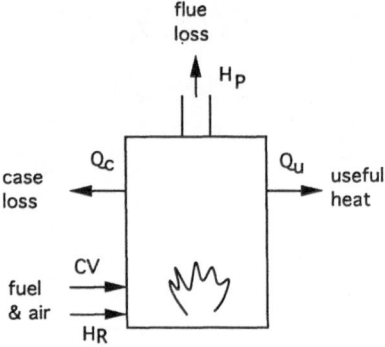

**Figure 4.1** Energy balance around a combustion system

considered. Before the energy balance is developed it is worth looking more closely at the terms involved:

(a) $H_R$, $H_P$ and CV are all related to the standard temperature of 25°C. The enthalpy of the entering air is considered to be sensible heat only (i.e. the fuel/air is considered to be a dry mixture).
(b) The enthalpy of the flue gas must be consistent with the calorific value of the fuel which is used in the balance. If the gross calorific value is used, then $H_R$ should contain a latent heat term equal to the mass of water produced per kilogram of fuel multiplied by the latent heat of evaporation of water at 25°C ($h_{fg}$). If the net calorific value is used, then the flue gas enthalpy will consist of sensible heat terms only. In this chapter we are concerned with predicting the temperature reached within the flame, hence the net calorific value/sensible heat terms system is the more appropriate.

The energy balance about the system can be written as:

$$CV + H_R = H_P + Q_c + Q_u \tag{4.1}$$

for 1 kg of fuel.

At this stage it is worth mentioning that for many purposes $H_R$, the sensible heat in the air and fuel (ref. 25°C) is very small and often neglected. The case loss from the outside of the plant, $Q_c$, is also generally small compared to the other energy fluxes and is similarly often considered negligible. In this situation the water vapour generated by the combustion reaction will remain in the vapour phase, hence CV represents the net calorific value of the fuel.

## 4.2 Adiabatic Flame Temperature

As a simplified picture of a sequence of events in the combustion of a fuel in a boiler, the first stage can be considered as adiabatic combustion of the fuel and air mixture, with the combustion products then entering a heat exchanger which extracts useful energy from the flame. This is a model which we will return to later in the book, but for now it is important to appreciate that the temperature of the flame as it enters the heat exchanger (Table 4.1) will have a significant effect upon the performance of the device. The higher the flame temperature, the greater should be the effectiveness of the heat exchanger section.

Table 4.1 Flame temperatures for some common fuels

| Fuel | Adiabatic flame temperature (°C) |
|---|---|
| Natural gas | 2070 |
| Kerosine | 2093 |
| Light fuel oil | 2104 |
| Medium fuel oil | 2101 |
| Heavy fuel oil | 2102 |
| Bituminous coal | 2172 |
| Anthracite | 2180 |

We can use the idea of the energy balance about the combustion system to derive a straightforward way of estimating the temperature of a flame. In this situation it is assumed that combustion takes place under adiabatic conditions, i.e. no heat transfer is permitted across the boundary of the system. The implication of this is that

$$Q_c = 0$$

and

$$Q_u = 0$$

hence equation 4.1 simplifies down to

$$CV + H_R = H_P \qquad (4.2)$$

In the development of this argument, it will be taken that CV is the net specific heat of the fuel, hence $H_P$ contains only sensible heat terms. The terms on the left-hand side of equation 4.2 are simple to evaluate as CV will be known for the fuel used and $H_R$, the enthalpy of the fuel and air (ref. 25°C), can easily be calculated from

$$H_R = (t_i - 25) \, \Sigma(m \, c_p)_R$$

where the summation $\Sigma$ is carried out for each of the species present in the reactants.

The specific heats of the fuel, oxygen and nitrogen can be evaluated at the mean temperature

$$(t_i + 25)/2$$

and the enthalpy of the reactants is thus easily evaluated. The enthalpy of the products, $H_P$, is then the sum of $H_R$ and the calorific value of the fuel, CV.

The right-hand side of equation 4.2, however, is not so easily evaluated as it is defined by

$$H_P = (t_f - 25) \, \Sigma(m \, c_p)_P \qquad (4.3)$$

and this relationship cannot be solved explicitly for $t_f$ as there will be a considerable difference between $t_f$ and the reference temperature 25°C, hence the value of $t_f$ is required to evaluate the specific heats of the combustion products.


# 4.3  Specific Heats of Gases

Simple ideal gas theory links an increase in the temperature of a gas to a corresponding increase in the kinetic energy of motion of the gas molecules. This model predicts that the specific heat of a gas is not a function of its temperature or pressure. While the latter implication is effectively true in practice, the specific heat of a gas does increase with temperature above about 100°C. This effect arises because molecules have vibrational (internal) energy as well as kinetic energy due to the motion of the complete molecule. This additional mode of 'storing' energy will mean that the specific heat of a gas will increase as its temperature rises. This cannot be modelled using the ideas of classical mechanics but the effect can be predicted using quantum theory.

Specific heats of gases are published in tabular form [1] but empirical equations which allow the calculation of specific heats as a function of temperature have been in use for some time [2]. A polynomial expression is normally used, with either fractional [3] or integer powers. With the widespread use of personal computers there is no real advantage in striving for a compact expression hence a straightforward integer power series is most convenient:

$$c_p = a[0] + a[1]t + a[2]t^2 + a[3]t^3 + \dots$$

**Table 4.2**  Polynomial coefficients for specific heats of gases

| Carbon Dioxide | Polynomial Order 4 |
|---|---|
| $a[0]$ | 0.818205 |
| $a[1]$ | $9.9739 \ e^{-4}$ |
| $a[2]$ | $-7.61047 \ e^{-7}$ |
| $a[3]$ | $2.79744 \ e^{-10}$ |
| $a[4]$ | $-3.8726 \ e^{-14}$ |
| | |
| Carbon Monoxide | Polynomial Order 4 |
| $a[0]$ | 1.030 |
| $a[1]$ | $0.1274 \ e^{-3}$ |
| $a[2]$ | $0.2414 \ e^{-6}$ |
| $a[3]$ | $-0.2174 \ e^{-9}$ |
| $a[4]$ | $0.4956 \ e^{-13}$ |
| | |
| Water (vapour) | Polynomial Order 4 |
| $a[0]$ | 1.86024 |
| $a[1]$ | $3.23229 \ e^{-4}$ |
| $a[2]$ | $5.84858 \ e^{-7}$ |
| $a[3]$ | $-3.5846 \ e^{-10}$ |
| $a[4]$ | $5.93307 \ e^{-14}$ |
| | |
| Oxygen | Polynomial Order 5 |
| $a[0]$ | 0.9057 |
| $a[1]$ | $0.2941 \ e^{-3}$ |
| $a[2]$ | $0.9650 \ e^{-7}$ |
| $a[3]$ | $-0.3364 \ e^{-9}$ |
| $a[4]$ | $0.2021 \ e^{-12}$ |
| $a[5]$ | $-0.3811 \ e^{-16}$ |
| | |
| Nitrogen | Polynomial Order 5 |
| $a[0]$ | 1.30709 |
| $a[1]$ | $8.11879 \ e^{-6}$ |
| $a[2]$ | $4.859 \ e^{-7}$ |
| $a[3]$ | $-4.61612 \ e^{-10}$ |
| $a[4]$ | $1.6814 \ e^{-13}$ |
| $a[5]$ | $-2.18231 \ e^{-17}$ |
| | |
| Hydrogen | Polynomial Order 5 |
| $a[0]$ | 14.350 |
| $a[1]$ | $0.9947 \ e^{-3}$ |
| $a[2]$ | $-0.2480 \ e^{-5}$ |
| $a[3]$ | $0.4978 \ e^{-8}$ |
| $a[4]$ | $-0.2856 \ e^{-11}$ |
| $a[5]$ | $0.5275 \ e^{-15}$ |

**Table 4.3**  Specific heats ($c_p$) of some gases (kJ kg$^{-1}$ K$^{-1}$)

| $t$ (°C) | $CO_2$ | $H_2O$ | $O_2$ | $N_2$ | CO |
|---|---|---|---|---|---|
| 0 | 0.818 | 1.860 | 0.906 | 1.037 | 1.030 |
| 25 | 0.843 | 1.869 | 0.913 | 1.038 | 1.033 |
| 50 | 0.866 | 1.878 | 0.921 | 1.039 | 1.037 |
| 100 | 0.911 | 1.898 | 0.936 | 1.042 | 1.045 |
| 150 | 0.952 | 1.921 | 0.951 | 1.048 | 1.054 |
| 200 | 0.989 | 1.946 | 0.966 | 1.055 | 1.063 |
| 250 | 1.024 | 1.972 | 0.981 | 1.063 | 1.074 |
| 300 | 1.056 | 2.001 | 0.995 | 1.072 | 1.084 |
| 350 | 1.086 | 2.031 | 1.009 | 1.082 | 1.096 |
| 400 | 1.112 | 2.062 | 1.022 | 1.093 | 1.107 |
| 450 | 1.137 | 2.094 | 1.035 | 1.104 | 1.118 |
| 500 | 1.159 | 2.127 | 1.046 | 1.115 | 1.130 |
| 550 | 1.180 | 2.161 | 1.057 | 1.126 | 1.141 |
| 600 | 1.198 | 2.195 | 1.067 | 1.137 | 1.153 |
| 650 | 1.215 | 2.230 | 1.077 | 1.148 | 1.164 |
| 700 | 1.230 | 2.264 | 1.086 | 1.159 | 1.175 |
| 750 | 1.244 | 2.300 | 1.094 | 1.170 | 1.185 |
| 800 | 1.256 | 2.334 | 1.101 | 1.180 | 1.195 |
| 850 | 1.268 | 2.368 | 1.107 | 1.190 | 1.205 |
| 900 | 1.278 | 2.403 | 1.113 | 1.199 | 1.214 |
| 950 | 1.287 | 2.436 | 1.119 | 1.208 | 1.223 |
| 1000 | 1.296 | 2.469 | 1.124 | 1.216 | 1.231 |
| 1050 | 1.303 | 2.502 | 1.128 | 1.224 | 1.238 |
| 1100 | 1.310 | 2.533 | 1.133 | 1.231 | 1.245 |
| 1150 | 1.316 | 2.564 | 1.137 | 1.237 | 1.252 |
| 1200 | 1.322 | 2.594 | 1.141 | 1.243 | 1.258 |
| 1250 | 1.328 | 2.623 | 1.144 | 1.249 | 1.263 |
| 1300 | 1.333 | 2.651 | 1.148 | 1.254. | 1.268 |
| 1350 | 1.337 | 2.678 | 1.151 | 1.259 | 1.272 |
| 1400 | 1.345 | 2.703 | 1.155 | 1.263 | 1.275 |
| 1450 | 1.346 | 2.728 | 1.159 | 1.267 | 1.279 |
| 1500 | 1.350 | 2.752 | 1.162 | 1.270 | 1.282 |
| 1550 | 1.354 | 2.774 | 1.166 | 1.273 | 1.284 |
| 1600 | 1.358 | 2.795 | 1.170 | 1.277 | 1.286 |
| 1650 | 1.361 | 2.815 | 1.174 | 1.279 | 1.288 |
| 1700 | 1.365 | 2.834 | 1.179 | 1.282 | 1.290 |
| 1750 | 1.369 | 2.852 | 1.183 | 1.284 | 1.292 |
| 1800 | 1.373 | 2.869 | 1.187 | 1.287 | 1.294 |
| 1850 | 1.376 | 2.885 | 1.192 | 1.289 | 1.296 |
| 1900 | 1.380 | 2.900 | 1.196 | 1.292 | 1.298 |
| 1950 | 1.383 | 2.914 | 1.199 | 1.294 | 1.301 |
| 2000 | 1.387 | 2.928 | 1.203 | 1.296 | 1.304 |

Over the range of temperatures of interest this method of evaluating specific heats at constant pressure will provide results of acceptable accuracy (around 0.1%). Values for the coefficients for the polynomials for some gases are given in Table 4.2. Exponential notation as understood by most procedural computer languages is used, i.e. $1.5\,e^{-4}$ means $1.5 \times 10^{-4}$. The power series returns a value for $c_p$ for the gas in kJ kg$^{-1}$ K$^{-1}$.

Polynomial series curve fits are a very convenient formulation to use but their chief limitation cannot be over-emphasised. Curve fits should *never* be used in the region outside the data set that was used to generate the fit. In the case of the above values, the data was obtained in the temperature range 25–2000°C and the polynomials should not be used outside these limits.

For hand calculations, tables of specific heats are a more useful formulation and Table 4.3 gives values for the common combustion gases over a typical range of temperatures for combustion in boilers. Straightforward linear interpolation should be used for intermediate temperatures.

# 4.4  Calculation Algorithm

There are several ways of solving implicit problems such the calculation of adiabatic flame temperature but a straightforward method is to execute the following steps:

1. evaluate the left-hand side of equation 4.2, i.e. $(CV + H_R)$;
2. guess a value for $t_f$ and use this to find the specific heats of the combustion products at the average between the flame and the reference temperature, i.e. $(t_f + 25)/2\,°C$;
3. solve equation 4.2 for $t_f$;
4. compare the new value of $t_f$ with the original estimate and if there is a substantial difference use the new value to re-evaluate the specific heats, looping back to 2 until satisfactory convergence is achieved.

With regard to step 2 above, taking the average temperature over the interval is only valid if the relationship between specific heat and temperature is linear. We have seen that this is not in fact the case and, to obtain an accurate value of the sensible heat in the combustion products above the reference temperature, an integrating calculation should be used. However, the accuracy of calculating the adiabatic flame temperature by this simple procedure is constrained by a number of limitations, as will be explained later, hence taking a simple arithmetical average is quite acceptable.

This method of calculating the flame temperature is, of course, the well-known technique of successive substitution which is used in many engineering calculations. While it is not a particularly efficient algorithm, it is stable and will generally converge to within 5°C in four iterations, which makes it suitable for hand calculation. The method is illustrated by the following example.

## Example 4.1

Find the adiabatic flame temperature for a stoichiometric methane/air flame if the initial temperature of the fuel and air is 10°C. Take the net calorific value of methane as 50.14 MJ kg$^{-1}$.

The first step is to evaluate, for 1 kg fuel, the mass of each of the reactants and products. These steps have been covered in previous chapters and the result can be summarised thus:

| Reactants | Products |
|---|---|
| 1 kg CH$_4$ | 2.75 kg CO$_2$ |
| 4 kg O$_2$ | 2.25 kg H$_2$O |
| 13.17 kg N$_2$ | 13.17 kg N$_2$ |

The initial temperature of the reactants is 10°C and the value of $H_R$ is the sensible heat in the reactants (ref. 25°C):

$$H_R = (10 - 25)\{(1.c_p CH_4) + (4.c_p O_2) + (13.17.c_p N_2)\}$$

The mean temperature of the reactants over the interval is 17.5°C. At this temperature the values of the specific heats are:

| | |
|---|---|
| CH$_4$ | 2.23 kJ kg$^{-1}$ K$^{-1}$ |
| O$_2$ | 0.92 |
| N$_2$ | 1.04 |

So         $H_R = -293.9$ kJ kg$^{-1}$ methane

Giving       $H_P = 50\,144 - 293.9$ kJ kg$^{-1}$
             $= 49\,850$ kJ kg$^{-1}$

This value of $H_P$ represents the sensible heat in the combustion products above the reference temperature of 25°C, i.e.

$$49\,850 = (t_f - 25)\{(13.17\,c_p N_2) + (2.25\,c_p H_2O) + (2.75\,c_p CO_2)\} \tag{4.4}$$

To solve this equation by successive substitution requires an initial guess for the flame temperature. As this procedure converges rapidly, the initial guess is not critical. If we assume 1575°C, the mean temperature of the products above 25°C is (1575 + 25)/2, i.e. 800°C.

The specific heats of nitrogen, water vapour and carbon dioxide at this temperature are

| | |
|---|---|
| N$_2$ | 1.18 kJ kg$^{-1}$ K$^{-1}$ |
| H$_2$O | 2.33 |
| CO$_2$ | 1.26 |

so the energy equation is

$$49\,850 = (t_f - 25)\{(13.17 \times 1.18) + (2.25 \times 2.33) + (2.75 \times 1.26)\}$$

$$t_f = 2081°C$$

The new mean temperature of the products is $(2081 + 25)/2 = 1053°C$.
   The specific heats are now:

| | |
|---|---|
| $N_2$ | 1.22 kJ kg$^{-1}$ K$^{-1}$ |
| $H_2O$ | 2.50 |
| $CO_2$ | 1.30 |

Solving the energy equation for the third time gives:

$$t_f = 1998°C$$

The mean temperature of the products now becomes $(1998 + 25)/2 = 1012°C$ giving

| | |
|---|---|
| $N_2$ | 1.22 kJ kg$^{-1}$ K$^{-1}$ |
| $H_2O$ | 2.48 |
| $CO_2$ | 1.30 |

   A further evaluation of the energy equation gives

$$t_f = 2001°C$$

There is no point in continuing the calculation beyond this point as the mean temperature above 25°C has now only changed by 1°C (to the nearest whole number). The values for the specific heats are only valid to two places of decimals at most, so the value of 2001°C represents the calculated adiabatic flame temperature for a stoichiometric methane/air mixture.

# 4.5  Calculated Adiabatic Flame Temperatures

Values of the predicted adiabatic flame temperature for hydrocarbon fuels are affected by the following factors:

1.   the calorific value (and chemical composition) of the fuel;
2.   the air-to-fuel ratio at which combustion takes place; and
3.   the initial temperature (preheat) of the air and fuel.

## Calorific Value

Clearly it would be expected that the higher the calorific value of the fuel the higher the final flame temperature produced. The picture is, however, not quite as simple as this since the chemical composition of the fuel also plays a part. As was shown in Chapter 3, fuels with a higher proportion of hydrogen have higher calorific values (on a mass basis); for example, natural gas has a carbon-to-hydrogen ratio of around 3.1 and a calorific value of 54 MJ kg$^{-1}$ (gross). A typical coal, on the other hand, has a carbon-to-hydrogen ratio of around 18.0 and a calorific value of 33.3 MJ kg$^{-1}$.

Fuels with a higher proportion of hydrogen will generate more water vapour in their combustion products; a glance at Table 4.2 shows that water vapour has a higher specific heat than the other combustion products, which will lower the adiabatic flame temperature.

The adiabatic flame temperatures for a number of common hydrocarbon fuels are given in Table 4.1. In each case combustion is stoichiometric and the initial temperature of the air and fuel is 25°C.

It can be seen that there is quite a small variation in the calculated flame temperature, given the wide variation between the calorific values of these fuels. This 'conformity' is due to the effect of the hydrogen in the fuel discussed above.

## Air-to-fuel Ratio

It is to be expected that the calculated flame temperature will be highest if combustion takes place with exactly the stoichiometric air requirement. Any excess air will increase the mass of flue gas relative to the mass of fuel, with a corresponding reduction in temperature.

With sub-stoichiometric air supply the flame temperature will also fall, as although the mass of flue gas is reduced, the effective calorific value of the fuel is also reduced by an amount equivalent to the calorific value of the carbon monoxide which is present in the flue gas. The net calorific value of carbon monoxide is 10.11 MJ kg$^{-1}$.

The graph of Fig. 4.2 shows the temperature calculated for a natural gas flame as a

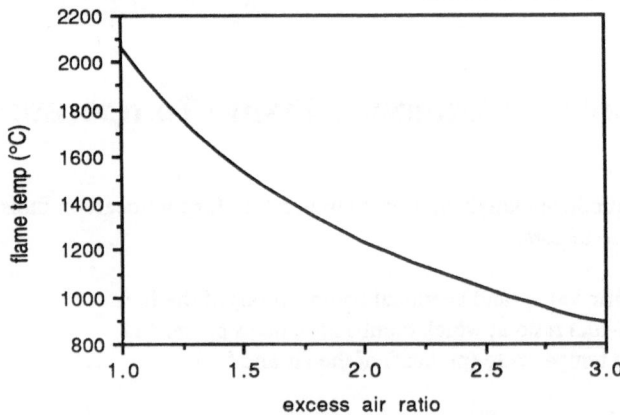

**Figure 4.2** Variation of natural gas flame temperature with stoichiometric ratio

function of the excess air ratio, and demonstrates the effects described above.

It is worth noting at this stage that very high flame temperatures can be achieved by oxygen enrichment of the flame, as the mass of flue gas is considerably reduced. Oxygen flames are used in a number of process applications, in particular welding, but the temperatures in oxygen-enriched flames are so high that dissociation effects (discussed later) are highly significant.

## Preheat

In many designs of burner the incoming air is preheated, usually by a heat exchanger which extracts low-grade heat from the flue gas. Common sense (and equation 4.2) suggest that preheating the air will increase the final temperature of the flame. This is a useful feature because by increasing the temperature difference between the flue gas and the heat extracting fluid, greater heat transfer will be obtained for the same fuel input rate. This will result in an increase in the thermal efficiency of the plant. An example of the effect of preheat on the adiabatic temperature of a natural gas flame at 5% excess air is shown in Fig. 4.3.

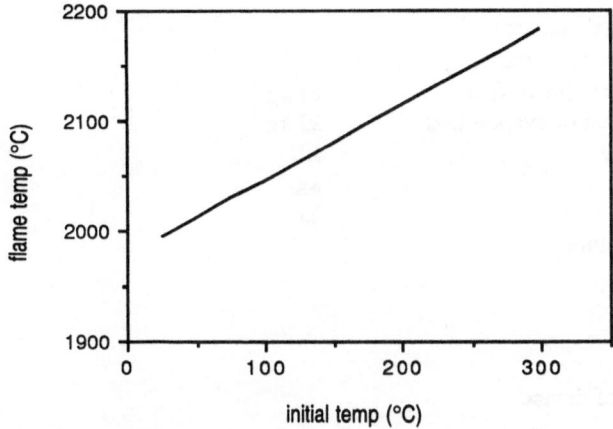

Figure 4.3 Effect of preheat on flame temperature

The flame temperatures calculated from the energy balance of equation 4.1 represent a simplification of what actually happens, and thus they represent an upper bound to the flame temperature. In practice, there are a number of reasons why the actual flame temperature will be lower than the value predicted by this procedure. The most significant of these is that at the high temperatures achieved in flames the combustion products will 'dissociate' back into reactants or other highly reactive species: this process is accompanied by an absorption of energy, hence reducing the actual flame temperature.

The procedure described above is very straightforward and its value lies in its suitability for hand calculation. The results of these calculations are more correctly referred to as 'undissociated adiabatic flame temperatures', which acknowledges the simplifications inherent in them. In order to understand the more complete picture it is necessary to gain some insight into the principles of chemical equilibrium, which are introduced in the next chapter.

## 4.6 References

1. Mayhew YR, Rogers GFC (1984). Thermodynamic and Transport Properties of Fluids. Blackwell, Oxford
2. Spencer HM (1945). Empirical heat capacity equations. J Amer Chem Soc 67: 1859–60
3. Stoecker WF (1989). Design of Thermal Systems, 3rd edn. McGraw-Hill, New York

## 4.7 List of Symbols

| | | |
|---|---|---|
| $a$ | numerical constant | |
| $c_p$ | specific heat (constant pressure) | kJ kg$^{-1}$ K$^{-1}$ |
| $CV$ | calorific value of fuel | kJ kg$^{-1}$ |
| $h_{fg}$ | latent heat of evaporation | kJ kg$^{-1}$ |
| $H$ | enthalpy | kJ |
| $m$ | mass | kg |
| $Q$ | heat | kJ |
| $t$ | temperature | °C |

*Subscripts*

| | |
|---|---|
| c | casing of device |
| f | flame |
| i | initial state |
| u | useful |
| P | products |
| R | reactants |

## 4.8 Problems

1. A distillate oil with an ultimate analysis of 85.7% carbon, 13.4% hydrogen has a gross calorific value of 45.5 MJ kg$^{-1}$. Calculate the adiabatic flame temperature if this fuel is burned at 15% excess air. The initial temperature of the air and fuel is 15°C.

   [1872°C]

2. A natural gas has the composition by volume

| methane | 93% |
| ethane | 4% |
| nitrogen | 3% |

and has a gross calorific value of 38.88 MJ m⁻³. Calculate the adiabatic flame temperature if this fuel is burned at 100% excess air (take the initial temperature of the air and fuel as 25°C).

[1232°C]

# Chapter 5

# Equilibrium Composition of Flames

## 5.1 Introduction

So far we have been able to treat the chemical changes taking place when a fuel is burned as one-way processes; in other words, the initial mixture of fuel and air burns completely to products (flue gas). For the standard design and analytical calculations this is a simple and adequate representation of what is taking place. However, there are circumstances under which the idea of a reaction proceeding to completion (i.e. from left to right in the conventional graphical notation) is not adequate to describe the situation. Examples of such situations are the formation of oxides of nitrogen and sulphur in flames and also in the accurate calculation of flame temperatures.

As a simple introduction to the idea of equilibrium composition we can look at the combustion of hydrogen to water vapour. This would normally be represented by the following equation:

$$H_2 + \tfrac{1}{2} O_2 \rightarrow H_2O$$

The implication of this expression is that after combustion the hydrogen and oxygen have completely combined to produce water. This reaction is *exothermic* – it produces a lot of heat. On the other hand, if water is subjected to very high temperatures it will dissociate, or split up, with the absorption of a considerable quantity of heat. This dissociation is termed an *endothermic* reaction. Although there are several reactions involved, the most significant dissociation reaction is simply the reverse of the above, i.e water is split up into molecular hydrogen and oxygen with the consequent absorption of heat:

$$H_2O \rightarrow H_2 + \tfrac{1}{2}O_2$$

It can now be seen that the one-directional combustion equation is, in fact, a simplification. As the reaction proceeds, the temperature of the water vapour produced in a flame will be around 2000°C which is sufficiently high to cause some dissociation of the water vapour in the flame. This means that the combustion products will contain some hydrogen and oxygen as well as the water vapour.

Dissociation has two important effects. Firstly, the flame temperature will be lower than that predicted by the simple adiabatic calculation described in Chapter 4. Secondly, the flame consists of dissociation products as well as the components predicted by the simple 'forward' reaction. We can go on to study this in more depth by looking at the equilibrium composition of flames.

It is essential to understand that an important limitation of this treatment is that it refers only to the equilibrium state, i.e. in which the system is stable and there are no property changes with time. There is a close parallel here with other systems analysed by classical thermodynamics, which is also concerned with equilibrium states.

## 5.2 Chemical Equilibrium

In the introductory section, two situations were considered, a forward reaction which is exothermic and a backward endothermic reaction. Any chemical reaction can be thought of as reversible, a state of dynamic equilibrium being achieved when the rates of the forward and backward reactions are equal.

This idea of equal reaction rates allows the use of a simple law to quantify the equilibrium composition of a reacting system. The Law of Mass Action states that the rate at which a substance reacts is proportional to its concentration. As we are concerned here with gases, the concentration of a gas is represented by its partial pressure.

If we consider a reacting system consisting of two reactants, A and B, combining to form two products, C and D we can write:

$$A + B \rightleftharpoons C + D$$

The rate of the forward reaction will, by the Law of Mass Action, be proportional to the concentration (partial pressure) of A $(p_A)$ and to the concentration of B, $p_B$. The forward reaction rate is thus proportional to the product of $p_A$ and $p_B$:

$$R_+ = k_+ p_A p_B$$

A similar argument for the backward reaction gives

$$R_- = k_- p_C p_D$$

At equilibrium the two rates are equal so $R_+ = R_-$ and dividing the two expressions above gives:

$$\frac{p_C p_D}{p_A p_B} = \frac{k_+}{k_-} = K \tag{5.1}$$

The term $K$ is called the *equilibrium constant* and is a function of the temperature of the system.

The number of molecules of each species, as expressed in the stoichiometric equation, affects the form of the expression for the equilibrium constant. If we have more than one of any molecule in the equation, for example

$$A + B \rightleftharpoons 2C$$

then the equilibrium constant will be

$$K = \frac{p_C^2}{p_A p_B}$$

as the rate of the backward reaction will be proportional to $(p_C p_C)$. Any chemical reaction can be expressed in terms of integer numbers of reacting species, but as has been noted earlier, it is often more convenient to use fractional numbers, as in

$$H_2 + \tfrac{1}{2}O_2 \rightleftharpoons H_2O \tag{5.2}$$

The equilibrium constant in this case is given by:

$$K_{5.3} = \frac{p_{H_2O}}{p_{H_2}\, p_{O_2}^{1/2}} \tag{5.3}$$

The way in which the equation is written is important, since if the combustion of hydrogen was written in integer form

$$2H_2 + O_2 \rightleftharpoons 2H_2O$$

the equilibrium constant would become

$$K_{5.4} = \frac{p_{H_2O}^2}{p_{H_2}^2\, p_{O_2}} \tag{5.4}$$

If the equilibrium values of the partial pressures of oxygen, hydrogen and water vapour, measured in the same units at the same temperature, were inserted into equations 5.3 and 5.4 then two different values for the equilibrium constant would be obtained. This shows that it is important to check the formulation of the equation when looking up numerical values for equilibrium constants.

In a gas mixture, the partial pressure of a component of the mixture is given by the total pressure ($p_T$) multiplied by the mole fraction (or volume fraction) of the constituent. If the total number of moles present in the mixture is $N$, equation 5.3 can be written:

$$K_{5.3} = \frac{n_{H_2O}}{n_{H_2}\,(n_{O_2}/N)^{0.5}}\, p_T^{-0.5} \tag{5.5}$$

There is a second dissociation reaction for water, significant at high temperatures, which produces the hydroxyl radical OH:

$$\tfrac{1}{2}H_2 + OH \rightleftharpoons H_2O \tag{5.6}$$

This reaction has its equilibrium constant defined by

$$K_{5.7} = \frac{p_{H_2O}}{p_{OH}\,(p_{H_2})^{0.5}} \tag{5.7}$$

The relative significance of this mechanism compared with equation 5.2 can be judged from the values of the respective dissociation constants given in Table 5.1.

For the dissociation of carbon dioxide

$$CO + \tfrac{1}{2}O_2 \rightleftharpoons CO_2 \tag{5.8}$$

the equilibrium constant is

$$K_{5.9} = \frac{p_{CO_2}}{p_{CO} \, (p_{O_2})^{0.5}}$$   (5.9)

or in terms of moles

$$K_{5.9} = \frac{n_{CO_2}}{n_{CO} \, (n_{O_2}/N)^{0.5}} \, p_T^{-0.5}$$   (5.10)

Values for the equilibrium constant for the above reactions are given in Table 5.1. High values for the equilibrium constant mean that the position of the equilibrium is over to the right, or 'product' side. Lower values of $K$ mean that the equilibrium is shifted correspondingly to the left. Tabulated values for the dissociation constant of water show, as would be expected, that high temperatures promote dissociation of the water vapour and that at low temperatures the equilibrium mixture consists almost entirely of water.

Table 5.1   Dissociation constants (partial pressures in atm) [1]

| Temp. (K) | $\log_{10} K_{5.3}$ ($H_2O$) | $\log_{10} K_{5.7}$ ($H_2O$) | $\log_{10} K_{5.9}$ ($CO_2$) |
|---|---|---|---|
| 300 | 39.79 | 46.29 | 44.74 |
| 400 | 29.24 | 33.91 | 32.41 |
| 600 | 18.63 | 21.47 | 20.07 |
| 800 | 13.29 | 15.22 | 13.90 |
| 1000 | 10.06 | 11.44 | 10.20 |
| 1200 | 7.896 | 8.922 | 7.742 |
| 1400 | 6.344 | 7.116 | 5.992 |
| 1600 | 5.175 | 5.758 | 4.684 |
| 1800 | 4.263 | 4.700 | 3.672 |
| 2000 | 3.531 | 3.852 | 2.863 |
| 2200 | 2.931 | 3.158 | 2.206 |
| 2400 | 2.429 | 2.578 | 1.662 |
| 2600 | 2.003 | 2.087 | 1.203 |
| 2800 | 1.638 | 1.670 | 0.807 |
| 3000 | 1.322 | 1.302 | 0.469 |

The behaviour of the position of such equilibria is a particular instance of a very useful fundamental rule (Le Chatelier's Principle) which states that when a system which is initially at equilibrium is put under any stress, the position of the equilibrium will move in a direction which will relieve the imposed stress. In this example, adding heat to the equilibrium mixture will drive the equilibrium position to the left, as the dissociation reaction is endothermic and this will 'absorb' the disturbance input.

When a hydrocarbon fuel is burned in air, the temperatures produced are limited by the presence of considerable quantities of nitrogen and by the excess air which is usually present. The significant dissociation reactions under these conditions are equations 5.2 and 5.8, so the gas mixture contains the molecular species carbon dioxide, carbon monoxide, hydrogen, water and nitrogen. Under equilibrium conditions the equilibrium constant relationships for both these reactions will be satisfied, but it can be seen that the oxygen present contributes to both relationships.

If the two equilibrium constant relationships 5.3 and 5.7 are divided we obtain

$$K_{5.11} = \frac{p_{CO} \, p_{H_2O}}{p_{CO_2} \, p_{H_2}} \tag{5.11}$$

or expressed in terms of number of moles

$$K_{5.11} = \frac{n_{CO} \, n_{H_2O}}{n_{CO_2} \, n_{H_2}} \tag{5.12}$$

This is known as the *water-gas* equilibrium, from the water-gas reaction:

$$CO_2 + H_2 \rightleftharpoons CO + H_2O$$

# 5.3  Calculation of the Equilibrium Composition

A quantitative example should help clarify the ideas expressed above. For this example, consider the stoichiometric combustion of hydrogen in air:

$$H_2 + \frac{1}{2}O_2 + \frac{1}{2}\left(\frac{79}{21}\right)N_2 \rightleftharpoons H_2O + \frac{1}{2}\left(\frac{79}{21}\right)N_2$$

or

$$H_2 + \tfrac{1}{2}O_2 + 1.881 \, N_2 \rightleftharpoons H_2O + 1.881 \, N_2$$

If we consider a simplified view of this system by limiting the 'backward' reaction to the dissociation of water as given by equation 5.2 it is convenient to introduce a variable $x$ representing the fraction of the product (that would be formed stoichiometrically) which has dissociated back into its original constituents. This means that the flue gas will contain

    $1 - x$ moles $H_2O$
    $x$     moles $H_2$
    $\dfrac{x}{2}$    moles $O_2$
    $1.881$   moles $N_2$

giving a total number of moles equal to

$$1 - x + x + \frac{x}{2} + 1.881$$

or $2.881 + \dfrac{x}{2}$ moles products.

We can now write expressions for the partial pressures of the reacting species in terms of the system total pressure $p_T$:

$$p_{H_2O} = \frac{(1-x)}{2.881 + \dfrac{x}{2}} p_T$$

$$p_{H_2} = \frac{x}{2.881 + \dfrac{x}{2}} p_T$$

$$p_{O_2} = \frac{\dfrac{x}{2}}{2.881 + \dfrac{x}{2}} p_T$$

The equilibrium constant for this reaction is:

$$K = \frac{p_{H_2O}}{p_{H_2} \, p_{O_2}^{0.5}}$$

Substituting the partial pressures above gives:

$$K = \frac{\dfrac{(1-x)}{\left(2.881 + \dfrac{x}{2}\right)}}{\dfrac{x}{\left(2.881 + \dfrac{x}{2}\right)} \dfrac{\left(\dfrac{x}{2}\right)^{0.5}}{\left(2.881 + \dfrac{x}{2}\right)}} p_T^{-0.5}$$

Manipulating this expression by squaring both sides and collecting terms in $x$ gives

$$x^3(1 - K^2 p_T) + 3.762x^2 - 10.524x + 5.762 = 0 \qquad (5.13)$$

A solution for this equation is required in the region $0 < x < 1$ and is quite easily found using trial and error. A computational algorithm can be easily written for this purpose if we note that if an estimate of $x$ is inserted into the left-hand side of equation 5.13, the value of the expression will not be zero but will be some 'residual' value. If the estimate of $x$ is lower than the value of $x$ at the solution point, the residual will be positive; but if the estimate is higher than the solution point value then the residual will be negative.

An algorithm for solving this problem is:

> set $x = 0$
> set an increment $d$ to an initial value (say 0.1)
>
> $x = x + d$
>> test residual value of the equation
>> repeat until the residual changes sign
> $x = x - d$

divide increment by 10 then repeat the above loop until the residual has reached a suitably low value

This algorithm is an implementation of the class of numerical methods known as the univariate search.

The calculation of the percentage dissociation of water and of carbon dioxide in stoichiometric mixtures at atmospheric pressure shows that carbon dioxide dissociates to a greater extent than water and that there is negligible dissociation below about 1700°C.

# 5.4 Dissociated Flame Temperature

The adiabatic flame temperature calculated by assuming that no dissociation occurs will clearly be higher than the true figure. Dissociation is an endothermic process, hence the energy absorbed by dissociation will come from the enthalpy of the gas mixture. The principles outlined above can be used as a basis for a more accurate flame temperature calculation using a simple iterative procedure.

The method is based on two calculation procedures: an equilibrium calculation as described above and a modified energy equation in which the calorific value of the fuel is reduced by the chemical energy of the hydrogen and carbon monoxide produced by the dissociation reactions.

The method starts by finding the undissociated adiabatic flame temperature, and then this temperature is used to obtain the equilibrium flue gas composition. The calorific value of the fuel is adjusted according to the amount of hydrogen and carbon monoxide produced and the energy equation solved for the current flue gas composition to give a recalculated flame temperature. This process is repeated until satisfactory convergence is achieved.

Such iterative procedures lend themselves very well to computational methods, but a manual example effectively illustrates the steps described.

## Example 5.1

Calculate the flame temperature for the stoichiometric combustion of hydrogen in air at atmospheric pressure. The initial temperature of the hydrogen and air is 25°C and the dissociation is to molecular hydrogen and oxygen (reaction 5.2).

*Stage 1: Adiabatic (Undissociated) Flame Temperature*

If the combustion reaction is assumed to go to completion we have

$$H_2 + \tfrac{1}{2}O_2 + \tfrac{1}{2} (3.76) N_2 \rightarrow H_2O + \tfrac{1}{2} (3.76) N_2$$

The net calorific value for the combustion of hydrogen is 241 800 kJ kmole$^{-1}$; with this type of problem it is generally easier to base the calculation on quantities expressed in kmoles, converting the specific heats (which are generally available on a mass basis) where appropriate.

As the initial temperature of the fuel and air is 25°C the energy equation is simplified to equating the calorific value of the fuel to the sensible heat in the flue gas relative to 25°C. The composition of the flue gas is:

| 1 kmole | (18 kg) | $H_2O$ |
| 1.881 kmole | (52.67 kg) | $N_2$ |

Guessing the flame temperature to be 1500°C and evaluating the specific heats at this temperature gives:

$$241\,800 = \{(18 \times 2.75) + (52.67 \times 1.27)\}\,(t_f - 25)$$

$$t_f = 2102°C$$

Two further iterations (in accordance with the procedure described in Chapter 4) give an undissociated adiabatic flame temperature of 2022°C.

*Stage 2: Equilibrium Composition*

Values of $K$ for the dissociation of water can be calculated from the following empirical expression [2]:

$$4.751 \log_{10} K = 57\,111 \times \frac{1}{T} - 2.6135 \log_{10} T - 0.8434 \times 10^{-3} T + 0.19602 \times 10^{-6} T^2 - 2.96716$$

(note that $T$ is the temperature in degrees Kelvin).

This expression gives a value of $K$ at 2022°C of 471.35. Equation 5.13 can now be solved to find the proportion of the stoichiometric water vapour which has dissociated:

$$(1 - 471.35^2)\,x^3 + 3.762\,x^2 - 10.524\,x + 5.762 = 0$$

whence $x = 0.029$. The flue gas now consists of:

| (1 − 0.029) | = | 0.971 kmoles $H_2O$ |
| | | 0.029 kmoles $H_2$ |
| 0.029/2 | = | 0.0145 kmoles $O_2$ |
| | | 1.881 kmoles $N_2$ |

and energy which can be quantified as unburned hydrogen with a calorific value of:

$$0.029 \times 241\,800 = 7012 \text{ kJ kmole}^{-1}\,H_2$$

The energy equation for the equilibrium mixture (where specific heats are available as kJ kmole$^{-1}$) can be written:

$$241\,800\,(1 - x) = (t_f - 25)\,\{c_p\,H_2O \times (1 - x) + x.\,c_p H_2 + \tfrac{1}{2}x.\,c_p O_2 + 1.881.\,c_p N_2\}$$

Solving this iteratively for $x = 0.029$ gives a revised flame temperature of 1973°C.

This value is then used to recalculate the equilibrium constant, the fraction of stoichiometric water vapour dissociated and the flame temperature. The 'outer loop' of this procedure converges typically after three iterations and the result is summarised below:

| Iteration | $x$ | $t_f$ (°C) |
|-----------|--------|------------|
| 1 | 0 | 2022 |
| 2 | 0.0291 | 1973 |
| 3 | 0.0241 | 1978 |
| 4 | 0.0246 | 1977 |

The adiabatic flame temperature, taking into account dissociation of water vapour, is thus 45°C lower than the undissociated value of 2022°C. This difference is about 2% and obviously careful consideration of the nature of the problem at hand is necessary before a flame temperature calculation taking into account dissociation is embarked upon. For most engineering purposes, an undissociated value for the flame temperature should suffice – the main reason for carrying out equilibrium composition calculations is when a detailed prediction of the composition of the combustion products is required.

The above calculation has been worked through to illustrate the principles concerned, solving the equilibrium composition equation(s) together with the energy equation. The calculation serves the additional purpose of giving a feel for the effect of dissociation on the calculated flame temperature. It contains, of course, many simplifications. Firstly, hydrogen is not a widely used fuel (it may have possibilities in powering vehicles). Secondly, writing the dissociation reaction as the reverse of the combustion reaction is a simplification as many more complex dissociation reactions occur. The dissociation of water itself occurs via two reactions instead of the single reaction considered in this example. The nitrogen and oxygen present may react to form oxides of nitrogen (this is important because of the emissions from plant although it has negligible effect on the flame temperature). Other reactions can take place, such as the splitting of molecular oxygen, hydrogen and nitrogen into atoms. Dissociation into atoms and radicals is, however, only significant at low pressures.

## 5.5  Dissociation with Hydrocarbon Fuels

The extension of the principles discussed above into the case of the combustion of a hydrocarbon fuel in air takes the problem outside the realm of what can be achieved with hand calculation. However, the numerical difficulties involved in solving the more general combustion system are easily resolved by computational means.

The combustion of a hydrocarbon fuel (C, H, O, N system) produces $CO_2$, $H_2O$, $O_2$ and $N_2$ as stoichiometric products. At temperatures above about 1700°C, dissociation reactions and the formation of nitric oxide (NO) become significant. The formation of oxides of nitrogen is important when considering emissions from thermal plant, although the calculated equilibrium concentrations are of limited significance.

The flue gas composition resulting from the combustion of a hydrocarbon fuel is described by a set of equilibrium reactions. The equilibria which follow are considered relevant here, although the treatment is general and can be applied to an extended or a reduced set of equations as desired:

$$CO_2 \rightleftharpoons CO + \frac{1}{2}O_2$$

$$H_2O \rightleftharpoons H_2 + \frac{1}{2}O_2$$

$$H_2O \rightleftharpoons OH + \frac{1}{2}H_2$$

$$\frac{1}{2}H_2 \rightleftharpoons H$$

$$\frac{1}{2}O_2 \rightleftharpoons O$$

$$\frac{1}{2}N_2 + \frac{1}{2}O_2 \rightleftharpoons NO$$

The variable $x$ (lower case) is used here to represent the unknown kmoles of each species. If we take a general hydrocarbon fuel containing $a$ kmoles of carbon and $b$ kmoles of hydrogen and assume that combustion takes place initially with $c$ kmoles of oxygen and $d$ ($= 3.76c$) kmoles of nitrogen, the combustion products will be

| | | |
|---|---|---|
| $x_1$ | kmoles | $CO_2$ |
| $x_2$ | kmoles | $CO$ |
| $x_3$ | kmoles | $H_2O$ |
| $x_4$ | kmoles | $H_2$ |
| $x_5$ | kmoles | $O_2$ |
| $x_6$ | kmoles | $OH$ |
| $x_7$ | kmoles | $N_2$ |
| $x_8$ | kmoles | $H$ |
| $x_9$ | kmoles | $O$ |
| $x_{10}$ | kmoles | $NO$ |

The total number of kmoles in this mixture, $N$, is given by:

$$N = x_1 + x_2 + x_3 + x_4 + x_5 + x_6 + x_7 + x_8 + x_9 + x_{10}$$

The equilibrium system is described by ten equations: the equilibrium relationships for the reactions above, together with the following continuity equations for carbon, hydrogen, oxygen and nitrogen.

Carbon balance:

$$x_1 + x_2 = a \tag{5.14}$$

Hydrogen balance:

$$x_3 + x_4 + 0.5(x_6 + x_8) = b \tag{5.15}$$

Oxygen balance:

$$x_1 + 0.5(x_2 + x_3 + x_6 + x_9 + x_{10}) + x_5 = c \qquad (5.16)$$

Nitrogen balance:

$$x_7 + 0.5x_{10} = d \qquad (5.17)$$

Equilibria:

CO$_2$/CO

$$\frac{x_1}{x_2(x_5/N)^{0.5}} \cdot p_T^{-0.5} = K_1 \qquad (5.18)$$

H$_2$O/O$_2$

$$\frac{x_3}{x_4(x_5/N)^{0.5}} \cdot p_T^{-0.5} = K_2 \qquad (5.19)$$

H$_2$O/OH

$$\frac{x_3}{x_6(x_4/N)^{0.5}} \cdot p_T^{-0.5} = K_3 \qquad (5.20)$$

H$_2$/H

$$\frac{x_8}{(x_4 N)^{0.5}} \cdot p_T^{-0.5} = K_4 \qquad (5.21)$$

O$_2$/O

$$\frac{x_9}{(x_5 N)^{0.5}} \cdot p_T^{-0.5} = K_5 \qquad (5.22)$$

NO/N$_2$

$$\frac{x_{10}}{(x_5 x_7)^{0.5}} \cdot p_T^{-0.5} = K_6 \qquad (5.23)$$

A computationally efficient method of solving general equation sets of this type is based on expressing each of the equations in residual form, that is with all the unknown terms on the left-hand side. The transformation of 5.14 into residual form is:

$$x_1 + x_2 - a = F_1$$

The value of $F_1$ at the solution point is, of course, zero as will be the value of all the residuals. The method requires that an initial estimate of the unknowns $x_1$ to $x_{10}$ is inserted into the equation set and the value of each of the residuals computed.

A search method is then used to refine the estimates for each of $x_1$ to $x_{10}$ such that the vector of residuals ($F_1$ to $F_{10}$) is forced to a near-zero value. The Newton method [3] is a fast and robust algorithm for the solution of this problem. The method has been shown to be a computationally efficient technique for the solution of this type of equation set [4] and is used in the flame thermodynamics computer program of Gordon and McBride [5].

Some results obtained from the solution of the above equation set for the stoichiometric combustion of methane in air at atmospheric pressure are shown in Table 5.2. The table shows mole fractions for each of the ten species within the temperature range 1400–2400 K. The dissociated flame temperature for this fuel is 2223 K. There is clearly no dissociation at the lower temperatures and at flame temperature there are only very small amounts of atomic hydrogen and oxygen.

Equilibrium calculations can become demanding of computer time if a large number of chemical reactions are to be considered. An instance of this is the formation of pollutants and corrosion-inducing substances in the combustion products from commercial liquid and solid fuels. If the scope of the problem is such that there are $n$ species in the combustion products, then in order to calculate the product composition and temperature (composition calculations on their own are of limited applicability) a set of $(n + 1)$ simultaneous equations must be solved.

The computer resources (memory and execution time) required to solve equation sets of this type increase roughly in proportion to $n^{2.5}$, hence hardware availability can become a significant issue for large systems of equations.

**Table 5.2** Equilibrium composition of methane combustion products

| | Temperature (K) | | | | | |
|---|---|---|---|---|---|---|
| | 1400 | 1600 | 1800 | 2000 | 2200 | 2400 |
| $N_2$ | 0.7149 | 0.7147 | 0.7142 | 0.7127 | 0.7092 | 0.7021 |
| $H_2O$ | 0.1901 | 0.1900 | 0.1894 | 0.1879 | 0.1843 | 0.1768 |
| $CO_2$ | 0.0950 | 0.0949 | 0.0941 | 0.0918 | 0.0862 | 0.0760 |
| CO | 0.0 | 0.00016 | 0.00088 | 0.00307 | 0.00820 | 0.0176 |
| $O_2$ | 0.0 | 0.00014 | 0.00052 | 0.00168 | 0.00428 | 0.00886 |
| $H_2$ | 0.0 | 0.00011 | 0.00045 | 0.00135 | 0.00330 | 0.00700 |
| OH | 0.0 | 0.00003 | 0.00018 | 0.00072 | 0.00223 | 0.00558 |
| NO | 0.0 | 0.00005 | 0.00021 | 0.00069 | 0.00179 | 0.00387 |
| O | 0.0 | 0.0 | 0.0 | 0.00003 | 0.00019 | 0.00088 |
| H | 0.0 | 0.0 | 0.00001 | 0.00006 | 0.00032 | 0.00133 |

# 5.6 References

1.  Mayhew YR, Rogers GFC (1984). Thermodynamic and Transport Properties of Fluids. Blackwell, Oxford
2.  Spiers HM (1962). Technical Data on Fuel. British National Committee, World Power Conference, London
3.  Rice JR (1983). Numerical Methods, Software, and Analysis. McGraw-Hill International, New York
4.  Prasad K (1970) Calculation of the equilibrium temperature and composition of a pre-mixed flame. Aeronautical Journal 74: 757–9
5.  Gordon S, McBride BJ (1971). NASA Report SP–273

# 5.7 List of Symbols

| | | |
|---|---|---|
| $c_p$ | specific heat (constant pressure) | kJ $(kg\ K)^{-1}$ |
| $d$ | numerical increment | – |

| | | |
|---|---|---|
| *F* | numerical residual | – |
| *k* | numerical constant | – |
| *K* | equilibrium constant | – |
| *n* | number of moles | – |
| *p* | partial pressure | atm |
| *R* | reaction rate | kmole s$^{-1}$ |
| *t* | temperature | degrees Celsius |
| *T* | temperature | degrees Kelvin |
| *x* | mole fraction | – |

*Subscript*

| | |
|---|---|
| T | total |

# Chapter 6

# Efficiency of Combustion Plant

---

## 6.1 Plant Efficiency

In Chapter 4 the idea of an energy balance around a combustion system was introduced and developed to give a simple procedure for calculating the adiabatic temperature of a flame. The general energy balance of the system was

$$CV + H_R = H_P + Q_c + Q_u \tag{6.1}$$

and this expression can be used to explore the meaning of efficiency in the context of heat-generating combustion plant.

Intuitively, the efficiency of a device such as a boiler would be interpreted as the rate at which useful heat is extracted from the device divided by the rate at which heat, in the form of the calorific value of the fuel being consumed, is supplied to the boiler. In practice, this does form the basis of the definition of efficiency of heating plant.

In the case of a device such as a boiler, the efficiency is defined as the rate of gain of enthalpy of the working fluid divided by the equivalent thermal energy input rate from the fuel. In the specific case, for instance, of a boiler supplying hot water, the instantaneous efficiency is given by:

$$E = \frac{m_w (t_2 - t_1)}{m_f \cdot CV} \tag{6.2}$$

At this point we must pause a moment and consider carefully what we mean by the calorific value of the fuel: do we use the net calorific value or the gross calorific value? In the majority of situations it does not matter provided that it is made quite clear which of the two calorific values is being used as a basis for the efficiency. If the net calorific value of the fuel is used we should refer to the 'net efficiency' of the boiler, if the gross calorific value is used then the figure obtained is called the 'gross efficiency'. As the net calorific value is lower than the gross figure then the net efficiency will have a higher numerical value.

The reason that the above paragraph was concerned only with the 'majority' of situations is that some boilers are designed to cool the flue gases to below their dew point (typically just below 60°C). These condensing boilers are thus capable of higher efficiencies as they are able to recover some of the latent heat present in the water vapour in the flue gas. As the net calorific value of the fuel excludes the latent heat of condensation of the water vapour, it is clearly appropriate to use the gross calorific value of the fuel in this context. If the flue gas is cooled significantly below the dew point of the water vapour present, then an efficiency figure greater than 100% can be obtained if the efficiency is based on the net calorific value. For this reason (and perhaps also for the feeling of unease that such numbers can induce) the efficiency of condensing devices should always be based on the gross calorific value of the fuel.

Before developing this discussion of efficiency of combustion appliances, it is worth taking a closer look at some of the terms in the steady-state energy balance of equation 6.1.

## (a)   Enthalpy of the reactants ($H_R$)

As the calorific value of the fuel represents the thermal energy available from the fuel based on the reference temperature of 25°C, the enthalpy of the fuel and air above or below this datum as they enter the system is a component of the energy balance. If the temperature of the air and fuel entering is close to the datum value, then clearly $H_R$ will not be a significant term in the balance. In other circumstances, such as when the fuel and air are preheated, then it will make a significant contribution to the equation.

It will be seen later, when describing practical measurements of combustion efficiency, that when the temperature of the entering fuel/air is other than 25°C, this can be partly compensated by basing the enthalpy in the combustion products with reference to the temperature of the fuel and air entering. This is acceptable provided that the thermal capacity rates (mass flow rate × specific heat of the mixture) of the fuel/air mixture entering the system and the combustion products leaving the system are approximately equal.

For most practical purposes, the combustion air is treated as a mixture of dry gases. The enthalpy of the combustion air above or below the reference temperature of 25°C is therefore given simply by the thermal capacities (mass × specific heat) of each of the components of the mixture multiplied by the difference in temperature between the mixture and 25°C. When developing a treatment of thermal efficiency in the condensing regime (Section 6.6), the presence of water vapour in the air entering is taken into account, and a latent heat is incorporated into the value of $H_R$ in this section.

## (b)   Thermal losses from the plant ($Q_c$)

This term is generally referred to as the 'case loss'. In simplified thermodynamic analyses the term is often neglected, as the heat transfer from the outer surfaces of a boiler is small compared to the throughput of the device. Clearly this term will be of decreasing significance as the size of the boiler increases, but a rough estimate of the figure is about 3–5% of the output of the boiler.

Case losses are normally an overall loss to the system, as boilers are generally not located in areas where the case loss can make a contribution to the heat load required. An obvious exception to this is the domestic central heating boiler, which is often located in a room such as a kitchen, where it can contribute to the heating of the building.

If any unburned fuel is present in the flue gas, this will be an energy loss from the system. An important objective of burner design is to obtain complete combustion of the fuel, hence good design and operation should keep this term to negligible proportions. Unburned stack losses can occur, for instance, in the combustion of liquid fuel sprays where particles of carbon can pass into the flue before they are completely burned.

By way of concluding this introductory section, it may be opportune to review the factors which affect the efficiency of a combustion appliance. These factors can be conveniently grouped if we regard a combustion appliance as a two-component system. The first component, the burner, has the task of converting the chemical energy in the fuel into thermal energy in the flame as efficiently as possible. To bring this about, there should be no unburned fuel in the flue gas and the fuel should be burned at an air-to-fuel ratio as close to the stoichiometric value as possible. Excess air merely passes through the system as a passenger and is warmed in the process, which inevitably leads to a reduction in the efficiency of the system.

There is generally an element of mutual exclusion here, as trimming the combustion air down to a value close to stoichiometric makes it more likely that imperfect mixing of the fuel and air will occur, leading to some unburned fuel leaving the system. The actual location of this compromise depends on the energy available for mixing the fuel and air streams. In a buoyancy-driven system such as a domestic gas-fired boiler this energy is low, hence large quantities of excess air are used to `avoid unburned fuel leaving the boiler (in this instance the unburned fuel will be in the form of carbon monoxide, so there are significant safety implications as well).

The second component of the system is a heat exchanger which transfers heat from the combustion products into a working fluid. The heat exchanger will thus have a gas on the combustion side of the exchanger and possibly a gas, liquid or boiling liquid on the other side of the exchanger. Of course, in practice these two components do not operate in a strictly sequential manner, as heat is generally extracted from the gases while combustion is still taking place, but this two-component model is a useful one which can easily be used as the basis for simple mathematical models.

At first sight it would seem that the heat exchanger should be as large (effective) as possible in order to extract maximum heat from the system. In fact, there are a number of constraints on the size of the exchanger that limit this. The effectiveness of a heat exchanger is primarily dependent on its overall coefficient of heat transfer and its heat transfer area (i.e. its size). The greater the surface area of the heat exchanger the higher will be its capital cost, and also the power required to push the combustion gases through it will increase. Increasing the heat transfer coefficient (by, for example, the use of turbulence promoters) will also increase the energy needed to maintain the flow through the device.

The last constraints that will be touched on here are concerned with the low gas temperatures which must accompany high heat recovery in the heat exchanger. The temperature of the heat-extracting fluid on entering must always be lower than the temperature of the flue gas on leaving. The implication of this is that very high boiler efficiencies can only be realised where there is a demand for low-grade heat. There are comparatively few applications where this is the case, although the heating of buildings is an important exception to this.

Finally, it is important to avoid the possibility of corrosion arising from unplanned condensation occurring within the combustion device, the heat exchanger or the flue. This consideration is particularly important with sulphur-bearing fuels, as any such condensate is acidic and therefore highly corrosive.

## 6.2  Direct Determination of Efficiency

The determination of the efficiency of combustion heating equipment by direct measurement is a task which is carried out in the laboratory. The basic measurements required are the mass flow rate, together with the temperatures of the heat-extracting fluid on entering (return) and leaving (flow), and the flow rate of the fuel supplied. The values of efficiency measured in this way are referred to as 'bench' efficiencies: they are instantaneous values which are normally higher than the efficiencies obtained over a period of time in an actual installation.

The value of efficiency obtained from such tests will depend on three factors: the fluid mass flow rate, the return temperature and the load on the device, which will determine its firing rate. A small device, such as a domestic or small commercial water-heating boiler, will usually have a single figure quoted for its efficiency. This figure will normally be obtained with a cool return temperature, a high water flow rate and a load (firing rate) at or near the appliance maximum. These conditions will produce the highest efficiency that the boiler is capable of.

Clearly the low return temperature together with a high flow rate will provide a low mean fluid temperature on the cool side of the heat exchanger, thus maximising the heat transfer rate. The effect of the load on the efficiency is a very important consideration, as this relationship will determine the efficiency of a device over a period of time in any given application. The control mode, by which the burner responds to sensed changes in load, is particularly significant here.

Most small (natural draught) boilers have a simple on/off control mode in which the burner is either firing at its full rate or is turned off. The hysteresis inherent in this control gives a variation of the temperature of the water in the boiler about the set point (Fig. 6.1). As the load on the boiler decreases, the burner is in the 'off' position

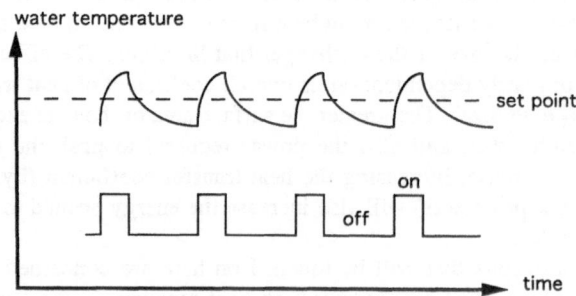

**Figure 6.1** On/off boiler operation

for an increasing proportion of the time. During these off periods, the loss from the casing continues, and there will also be heat transfer in the reverse direction from the hot working fluid to the gas (in many cases, air-induced through the boiler by buoyancy forces) in the heat exchanger, which further reduces the efficiency of energy conversion.

The variation of bench efficiency with load is shown in Fig. 6.2 for a typical natural draught domestic boiler (on/off control) and a larger unit of commercial size (continuous modulation of air and fuel) [1]. It can be seen that the efficiency falls off markedly at low loads, so that if there is a significant variation in the load pattern careful thought should be given to the specification of multiple boilers and their load scheduling in order to avoid low-load operation.

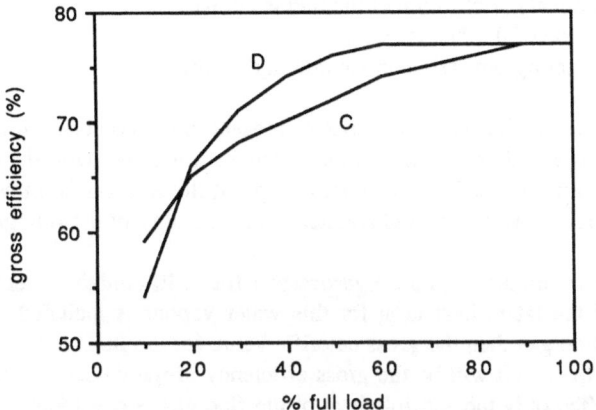

**Figure 6.2** Efficiency curves for typical domestic (D) and commercial (C) boilers (after [1])

# 6.3  Inferential Measurement of Efficiency

The direct measurement of boiler, or appliance, efficiency clearly requires intrusive instrumentation and careful control of operating conditions. There are, however, frequent occasions when the determination of the operating efficiency of a heating device can yield information which is valuable in making decisions regarding maintenance, replacement or commissioning of combustion equipment. Compact, portable combustion analysers form an essential part of the equipment of any engineer concerned with thermal combustion systems and, although these instruments are of ever-increasing sophistication in their measurement capabilities, the principle of the analysis of the efficiency of the combustion device is general and straightforward.

Inferential measurement is based on readings which enable the enthalpy losses in the flue gas (per unit mass of fuel) to be related to the calorific value of that fuel. Rearranging equation 6.1, and assuming that $Q_c$ is negligible, we get:

$$CV = (H_P - H_R) + Q_u$$

The term in brackets represents the difference between the enthalpy of the air/fuel and that of the combustion products, measured from the reference temperature of 25°C.

Given that the efficiency of the system is defined by $Q_u/CV$, we can write:

$$E = 1 - \frac{(H_P - H_R)}{CV} \qquad (6.3)$$

Knowing the calorific value of the fuel, the values of $H_P$ and $H_R$ now need to be determined. In order to evaluate these, it is necessary to know the temperature and composition of the flue gas and of the fuel/air mixture entering.

If the following conditions apply:

> The chemical composition of the fuel is known
> Combustion of the fuel is complete
> There is negligible carbon monoxide in the products

then a measurement of any *one* of the species present in the combustion products will enable the air-to-fuel ratio on entering to be calculated, together with the composition of the combustion products. Once the composition of the flue gas is known, the mass of each constituent is easily obtained and hence the enthalpy of the flue gas above the required datum.

The combustion products from a hydrocarbon fuel will contain a certain mass of water vapour. If the latent heat term for this water vapour is included in the (total) enthalpy of the flue gas then the gross calorific value for the fuel must be used in the calculation, and the result will be the gross efficiency of the device. If the latent heat term is omitted (i.e. only the sensible heat of the flue gas is considered) then the net calorific value must be used and the resulting figure will be the net efficiency.

Contemporary combustion analysers give a reading of either $\%CO_2$ or $\%O_2$ in the flue gas. These are invariably percentages of the dry combustion products (strictly speaking, the gas they analyse will be saturated with water vapour). The analyser has a single thermocouple probe which, if the scale is zeroed to the ambient temperature, will provide a direct estimate of $(H_P - H_R)$, assuming that there is little difference between the thermal capacity of the products and the air-to-fuel mixture. Example 6.1 illustrates the principle of working out a boiler efficiency from these simple experimental readings.

## Example 6.1

A propane-fired boiler is operating in a plant room where the ambient temperature is 15°C. Measurements at the exit from the heat exchanger showed a flue gas temperature of 240°C and a reading of 9% $O_2$. Estimate the gross and net efficiencies of the boiler. Take the net calorific value of propane as 46.48 MJ kg$^{-1}$.

In this calculation it will be necessary to work in both volumetric and mass units. The stoichiometric equation is an appropriate starting point:

$$C_3H_8 + 5O_2 + ...N_2 \rightarrow 3CO_2 + 4H_2O + ...N_2$$

| 1 vol | 5 | 16.81 | 3 | 4 | 16.81 |
|-------|------|-------|---|-------|-------|
| 1 kg | 3.636 | 11.97 | 3 | 1.636 | 11.97 |

This stoichiometric equation does not include the excess air; this is easily established from the $\%O_2$ reading, provided there is complete combustion and no CO present.

The combustion products contain

    3      vols $CO_2$
16.81   vols $N_2$

$x$     vols   excess air, consisting of
0.21$x$ vols   $O_2$ and
0.79$x$ vols   (excess) $N_2$

so if there is 9% oxygen in the products

$$\frac{9}{100} = \frac{0.21x}{x+3+16.81}$$

giving $x = 16.36$ vols air.
    The stoichiometric air requirement is

$$\frac{5}{0.21} = 23.81 \text{ volumes,}$$

so the excess air is

$$\frac{16.36}{23.81} \times 100\% = 66.7\%.$$

The quantities of reactants and products can now be established on a mass basis:

*Reactants:*

|  |  |  |
|---|---|---|
| 1 | kg | $C_3H_8$ |
| 3.636 | kg | stoichiometric oxygen |
| 2.50 | kg | excess oxygen |
| 20.19 | kg | nitrogen |

*Products:*

|  |  |  |
|---|---|---|
| 3 | kg | $CO_2$ |
| 1.636 | kg | $H_2O$ |
| 2.50 | kg | $O_2$ |
| 20.19 | kg | $N_2$ |

The next step is to establish the enthalpies of the reactants and products from the datum temperature of 25°C. At this stage the latent heat terms will be omitted.

*Reactants (mean temperature 25°C):*

Specific heats:

|  |  |
|---|---|
| $C_3H_8$ | 1.97 kJ (kg K)$^{-1}$ |
| $O_2$ | 0.91 |
| $N_2$ | 1.04 |

$$H_R = (15 - 25)\{(1 \times 1.97) + (6.0 \times 0.91) + (20.19 \times 1.04)\}$$

$$= -302.5 \text{ kJ kg}^{-1} \text{ of fuel}$$

*Products (mean temperature 133°C):*

Specific heats:

$$
\begin{array}{ll}
CO_2 & 0.94 \ \text{kJ (kg K)}^{-1} \\
H_2O & 1.91 \\
O_2 & 0.95 \\
N_2 & 1.05
\end{array}
$$

$$CV = (240 - 25)\{(3.0 \times 0.94) + (1.636 \times 1.91) + (2.5 \times 0.95) + (20.19 \times 1.05)\}$$

$$= 6346.6 \ \text{kJ kg}^{-1} \ \text{of fuel}$$

$$H_P - H_R = 6346.6 + 302.5$$

$$= 6649 \ \text{kJ kg}^{-1} \ \text{of fuel}$$

At this point it might be instructive to investigate the error which would have resulted if, as happens in many practical situations, the probe used to obtain the flue gas temperature had been zeroed to the ambient temperature of 15°C. The value for $(H_P - H_R)$ would then have been calculated from the datum of 15°C for the gas composition measured in the flue.

The mean temperature over the interval $(240 - 25)$ is 113°C and the specific heats at this temperature are:

$$
\begin{array}{ll}
CO_2 & 0.92 \ \text{kJ (kg K)}^{-1} \\
H_2O & 1.90 \\
O_2 & 0.94 \\
N_2 & 1.04
\end{array}
$$

Hence:

$$H_P - H_R = (240 - 15)\{(3.0 \times 0.92) + (1.636 \times 1.90) + (2.5 \times 0.94) + (20.19 \times 1.04)\}$$

$$= 6574 \ \text{kJ kg}^{-1} \ \text{of fuel}$$

The error resulting from this assumption, which is due to the thermal capacities (mass × specific heat) of the constituents of the flue gas being different from those of the reacting air and fuel, amounts to around 1% in this case.

The net efficiency of the boiler can now be calculated from:

$$E_n = \left(1 - \frac{(H_P - H_R)}{CV_n}\right) \times 100\%$$

Remembering that the above calculation refers to sensible heat terms only, the net efficiency of the boiler is:

$$E_n = \left(1 - \frac{6649}{46\,480}\right) \times 100\%$$

$$= 85.7\%$$

To get the gross efficiency we must add a latent heat term equal to the mass of water vapour produced (1.636 kg per kg fuel) multiplied by the latent heat of evaporation of water at 25°C (2442 kJ kg$^{-1}$), i.e. 3995 kJ.

The term $(H_P - H_R)$ now becomes (6649 + 3995) = 10 644 kJ, and the gross calorific value of the fuel is (46 480 + 3995) = 50 475 kJ kg$^{-1}$.

The gross efficiency of the boiler is thus:

$$E_g = \left(1 - \frac{10644}{50475}\right) \times 100\%$$

$$= 76.9\%$$

## 6.4 Efficiency and Flue Gas Temperature

The temperature of the flue gas as it leaves the combustion device clearly determines its thermal efficiency; neglecting case losses, the lower the temperature of the combustion products on leaving, the greater will be the thermal efficiency of the system. In the above example, a gross thermal efficiency of 77% resulted from a flue gas leaving at a temperature of 240°C. If a larger heat exchanger were to be fitted then more useful heat could be extracted from the flue gas and the efficiency would increase: a relationship between efficiency and flue gas temperature could easily be established for any given fuel and excess air level.

Figure 6.3 shows the relationship between gross efficiency and flue gas temperature for the combustion of a propane-fired heating device operating with 25% excess air. For convenience, the air/fuel temperature on entering has been taken as 25°C.

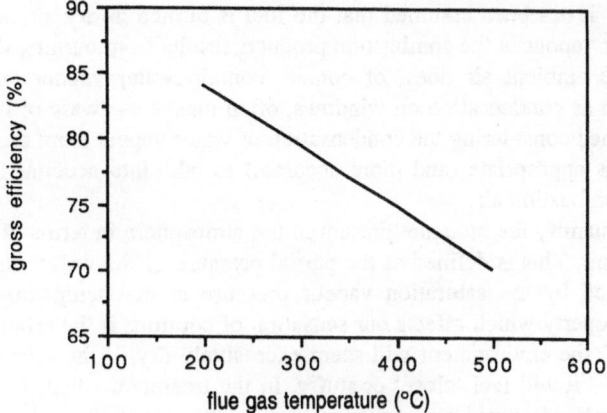

**Figure 6.3** Effect of flue gas temperature on efficiency

Cooling the flue gas to even lower temperatures is feasible, if there is a demand for low-grade heat (as occurs, for instance, in the heating of buildings). As the temperature is reduced, there will come a point at which the flue gas is below its dew point and condensation of the water vapour produced by the combustion of the fuel will take place. Devices which operate in this regime are referred to as 'condensing boilers' and they are capable of very high thermal efficiencies.

## 6.5   Flue Gas Dew Point

Before an analysis of the relationship between flue gas temperature and efficiency can be extended into the condensing regime, the dew point (the temperature at which water begins to condense out of the gas mixture) of the flue gases must be considered. One of the simplest ways of finding the dew point temperature of the flue gases is based on the volumetric percentage of water vapour in the combustion products. As a consequence of the ideal gas laws, if $x\%$ of the molecules in a gas mixture are of a given species (e.g. water vapour), then they will exert $x\%$ of the total pressure of the mixture. In other words, the partial pressure of this constituent will be $x\%$ of the total pressure of the mixture. This means, for example, that if the wet flue gas composition shows 10% water vapour and the flue gas is at atmospheric pressure, then the partial pressure of the water vapour is 0.1 atm.

The water vapour will start to condense when the temperature falls to such a level that its partial pressure is equal to the saturation vapour pressu.e at that temperature. The relationship between saturation vapour pressure and temperature is generally obtained from steam tables [2].

Alternatively, the saturation vapour pressure (in kPa) can be calculated from [3]:

$$\log_{10} p_s = 30.59051 - 6.2 \log_{10} T + 2.4804 \times 10^{-3} T - (3142.31/T) \qquad (6.4)$$

where $T$ is the temperature in degrees Kelvin.

Up until now it has been assumed that the fuel is burned in dry air, which means that all the water vapour in the combustion products results from burning the hydrogen in the fuel. The ambient air does, of course, contain water vapour and everyday experience, such as condensation on windows, often makes us aware of its dew point temperature. When considering the condensation of water vapour from the products of combustion, it is appropriate (and more accurate) to take into account the moisture present in the combustion air.

We usually quantify the moisture present in the atmosphere in terms of the *relative humidity* of the air. This is defined as the partial pressure of the water vapour present in the air divided by the saturation vapour pressure at that temperature. Relative humidity is a property which affects our sensation of comfort; if the relative humidity is less than 30% the environment will seem excessively dry, if the relative humidity rises to over 70% it will feel 'close' or stuffy. In the treatment which follows, use is made of a property of humid air known as the *moisture content* (or *humidity ratio*). This is frequently given the symbol $g$ and is defined as the mass of water vapour (in kg) associated with 1 kg of dry air. It is a useful property in the current context as it enables the mass of water vapour contributed to the flue gas by the humidity in the air to be calculated.

The moisture content at saturation can be calculated, with adequate accuracy for this task, from [3]:

$$g_s = \frac{0.624 p_s}{p_{atm} - 1.004 p_s}$$

If the gas mixture is at standard atmospheric pressure (1.01325 bar) this expression becomes:

$$g_s = \frac{0.624 p_s}{1.01325 - 1.004 p_s} \tag{6.5}$$

For example, at 15°C the saturation vapour pressure of air is 0.017 04 bar. Substitution of this value into equation 6.5 gives a saturation moisture content of 0.017 kg water kg$^{-1}$ dry air. A typical value for the relative humidity of the ambient air is 60%. Air at 60% relative humidity at a temperature of 15°C will have a moisture content of about 0.01 kg kg$^{-1}$. It will be assumed in the numerical working below that this figure is representative of the combustion air.

## Example 6.2

As an illustration of this method, we can estimate the dew point of the flue gas from the combustion of propane with 25% excess air, as in the example boiler efficiency calculation above.

Without reiterating the stoichiometric calculations, the combustion of 1 kg of propane requires 15.6 kg air (taken to be dry). At a moisture content of 0.01 kg kg$^{-1}$ dry air, this quantity of air will contain 0.156 kg water vapour. If the fuel is burned with 25% excess air then 0.195 kg (about 0.2 kg) water vapour will appear in the flue gas from the air supply. The total quantity of water vapour in the flue gas is thus:

| | |
|---|---|
| From combustion of the fuel | 1.636 kg |
| From the air supply | 0.195 kg |
| *Total* | 1.831 kg |

The combustion of 1 kg propane produces:

| | | |
|---|---|---|
| 3.0 | kg | $CO_2$ |
| 1.831 | kg | $H_2O$ |
| 0.909 | kg | $O_2$ |
| 14.96 | kg | $N_2$ |

We need to find the volume fraction of water vapour in these products; this will be given by the mole fraction of water vapour. The number of moles of each of the constituents is:

$CO_2$   (3.0/44)   0.068
$H_2O$   (1.836/18)   0.102
$O_2$   (0.909/32)   0.028
$N_2$   (14.96/28)   0.534
_____

*Total*   0.732

The mole fraction of water vapour is thus $0.102/0.732 = 0.139$. The partial pressure of water vapour in the flue gas is thus $0.139 \times 1.10325 = 0.141$ bar. From steam tables, the corresponding saturation temperature is 52.6°C.

The dew point of the combustion products is therefore *52.6°C.*

# 6.6  Efficiency of a Condensing Boiler

The significance of the dew point, in terms of boiler efficiency, is that if the combustion products are cooled below this temperature, the latent heat of condensation will be extracted in addition to the sensible heat of cooling of the products of combustion. The efficiency of a boiler in the condensing regime can be calculated in a straightforward manner from the energy equation by including an appropriate latent term in the energy equation:

$$E = 1 - (CV - H_R)/CV$$

Here, an arbitrary temperature datum of 25°C has been assumed for CV and $H_R$, as this temperature is the reference at which the calorific values of fuels are reported. This leads to the following expression for the sensible heat component of the enthalpy of the reactants:

$$\Sigma(m \times c_p)_R\ (t - 25)$$

The presence of water vapour can be accounted for in thermodynamic terms by adding to this a latent heat component equal to the mass of water vapour present multiplied by the latent heat of evaporation of water at 25°C (2442 kJ kg$^{-1}$). This gives an expression for the evaluation of the total (sensible + latent) enthalpy of a gas mixture

$$H = \Sigma(m \times c_p)\ (t - 25) + 2442 m_{H_2O} \qquad (6.6)$$

which is applicable to both reactants and products.

This expression has already been utilised in the evaluation of gross boiler efficiency of Example 6.1. As combustion in dry air was assumed in this case, all the water vapour in the products came from the combustion of the hydrogen in the fuel. If the combustion products are cooled below the dew point, then some of this water vapour will have been condensed out and it is necessary to determine the quantity of water vapour remaining at the product temperature to enable $H_R$ to be evaluated.

The saturation vapour pressure $p_s$ can be obtained from steam tables or equation 6.4, but it is necessary to find the corresponding mass of water vapour in the gas mixture. Assuming that, as a result of a stoichiometric calculation, the masses of each of the constituent reactants and products are available per kg of fuel burned, the number of moles of each constituent is easily obtained. Assuming that we have $n$ moles of water vapour and $N$ moles of dry combustion products and that the system total pressure is $p$ then at the saturation condition, s, we can write the following expression for the saturation vapour pressure, $p_s$

$$\frac{p_s}{p} = \frac{n_s}{n_s + N}$$

which is rearranged to give

$$n_s = \frac{p_s}{p - p_s} N$$

hence the mass of water vapour at saturation present in the products is

$$m_s = \frac{18 p_s}{p - p_s} N \qquad (6.7)$$

The effect of operating a boiler in the condensing regime can be illustrated by recalculating Example 6.1 assuming that the flue gas is cooled below the dew point.

## Example 6.3

What is the efficiency of a propane-fired boiler operating with 25% excess air if the combustion products are cooled to 40°C?

The stoichiometric quantities have already been evaluated in the previous examples and so their values will be used again here. The first step is to find the mass of water vapour present at saturation in the flue gas at 40°C, as this will enable the amount of water vapour condensed out to be obtained.

The dry combustion products are:

|       | mass (kg) | mol. wt. | moles  |
|-------|-----------|----------|--------|
| $CO_2$ | 3.0       | 44       | 0.0682 |
| $O_2$  | 0.909     | 32       | 0.0284 |
| $N_2$  | 14.96     | 28       | 0.5343 |
| Total ($N$) |      |          | 0.6309 |

At 40°C the saturation vapour pressure of water (from steam tables) is 0.073 75 atm. The corresponding mass of water vapour in the combustion products is, from equation 6.7:

$$m_{H_2O} = \frac{0.07375}{1.01325 - 0.07375} \times 18 \times 0.6309$$

$$= 0.891 \text{ kg}$$

The total mass of water produced is 1.831 kg (see above) and the specific heats of the gases and vapours are:

$CO_2$   0.86
$H_2O$   1.87
$O_2$    0.91
$N_2$    1.04

The enthalpy of the products, including the latent heat term, from the datum of 25°C is:

$$CV = (40 - 25)\{(3 \times 0.86) + (1.831 \times 1.87) + (0.909 \times 0.91) + (14.96 \times 1.04)\} + (0.891 \times 2442)$$
$$= 2511 \text{ kJ kg}^{-1} \text{ fuel}$$

Similarly, the enthalpy of the reactants with respect to the same datum is:

$$H_R = (15 - 25)\{(1 \times 1.97) + (0.1951 \times 1.87) + (4.545 \times 0.91) + (14.96 \times 1.04)\} + (0.1951 \times 2442)$$

$$= 256 \text{ kJ kg}^{-1} \text{ fuel}$$

Hence:
$$H_R - CV = 2511 - 256$$
$$= 2255 \text{ kJ kg}^{-1}$$

The gross calorific value of propane is 50 475 kJ kg$^{-1}$, hence the efficiency is:

$$E = 1 - \frac{2255}{50745}$$
$$= 95.5\%$$

The curve of gross efficiency of this system as a function of product temperature is shown in Fig. 6.4; this form of efficiency curve is a typical shape for a condensing device. There shows the sharp discontinuity at the combustion product's dew point as

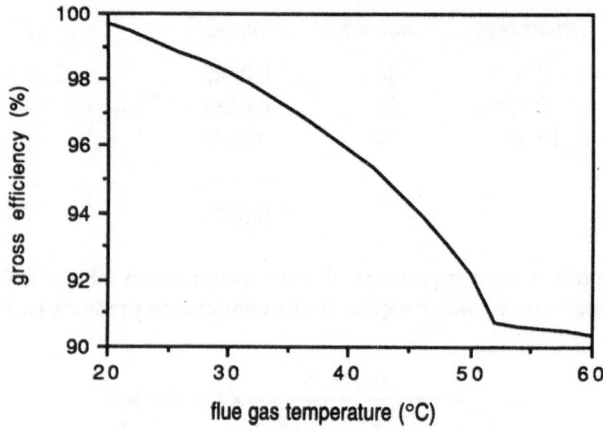

**Figure 6.4** Efficiency of a condensing boiler

condensation begins to occur, and then the curve becomes asymptotic at very low temperatures.

The reader may wonder why the gross efficiency of the system is not exactly 100% when the combustion products are cooled to the reference temperature of 25°C. To all practical intents, this is the case. However, the discrepancy is due to the capacity of the combustion air and combustion products to hold water vapour. In this example the combustion air contained only 60% of the water vapour which it was capable of holding, and cooling the combustion products down to 25°C does not condense out all the water vapour which was produced by combustion of the fuel, although the latent heat associated with this water vapour appears explicitly in the gross calorific value of the fuel.

The higher the excess air at which the fuel is burned, the greater will be the capacity of the combustion products to hold water vapour and the higher will be this small 'efficiency gap'.

# 6.7 References

1. Letherman KF, Dewsbury J (1986). The 'bin' method – a procedure for predicting seasonal energy requirements for buildings. Building Services Engineering Research and Technology 7: 55–64
2. Mayhew YR, Rogers GFC (1984). Thermodynamic and Transport Properties of Fluids. Blackwell, Oxford
3. CIBSE Guide (1986). Chartered Institution of Building Services Engineers, London

# 6.8 List of Symbols

| $c_p$ | specific heat (constant pressure) | kJ kg$^{-1}$ |
| $CV$ | calorific value | kJ kg$^{-1}$; kJ m$^{-3}$ |
| $E$ | efficiency | – |
| $g$ | moisture content | kg kg$^{-1}$ |
| $H$ | enthalpy | kJ (kg fuel)$^{-1}$ |
| $m$ | mass | kg |
| $m$ | mass flow rate | kg s$^{-1}$ |
| $p$ | partial pressure | atmospheres |
| $Q$ | heat | kJ (kg fuel)$^{-1}$ |
| $t$ | temperature | degrees Celsius |
| $T$ | temperature | degrees Kelvin |

*Subscripts*

atm      atmospheric pressure

| c | casing of device |
|---|---|
| f | fuel |
| g | gross |
| n | net |
| s | saturation |
| T | total |
| w | water |

| 1 | inlet conditions |
|---|---|
| 2 | outlet conditions |

## 6.9 Problems

1. A probe monitoring the percentage by volume of oxygen in the flue of a propane-fired boiler showed a reading of 4% (dry). What would be the expected reading for carbon dioxide, assuming complete combustion was taking place?

   [11.1%]

2. An oil consisting of 87% carbon and 13% hydrogen (by mass) is burned at atmospheric pressure with 15% excess air. Calculate the dew point of the products, assuming that the combustion air is dry.

   [47°C]

3. The flue gas leaving the heat exchanger of an oil-fired boiler was measured at 325°C with 9% $CO_2$. Estimate the gross and net operating efficiencies. The oil consists of 86% carbon and 14% hydrogen (by mass) and has a gross calorific value of 45.5 MJ kg$^{-1}$.

   [75.4%, 80.9%]

4. A refuse-fired boiler is used to provide heat for a district heating scheme. The refuse input rate is 280 tonnes per day and the net calorific value of the fuel is 11.9 MJ kg$^{-1}$. The rate of extraction of useful heat from the boiler is 25.5 MW.

   The flue gas temperature leaving the boiler is at 390°C and the volumetric composition of the flue gas is as follows:

   | $CO_2$ | 8% |
   |---|---|
   | $H_2O$ | 11% |
   | $O_2$ | 5% |
   | $N_2$ | 76% |

   Estimate the mass flow rate of the flue gas.

[32.2 kg s$^{-1}$]

5. A gaseous fuel has the following composition by mass:

| | |
|---|---|
| carbon | 75% |
| hydrogen | 23% |
| nitrogen | 2% |

The gross calorific value of the gas is 53.97 MJ kg$^{-1}$ and its specific heat at constant pressure is 2.22 kJ kg$^{-1}$. Estimate the dew point of the combustion products of this fuel and the efficiency if it burned in a boiler at 40% excess air with a flue gas temperature of 40°C. The initial temperature of the air and fuel is 20°C.

[52°C, 95.5%]

# Chapter 7

# Gaseous Fuels

## 7.1 Introduction

Hydrocarbon fuels are burned in gas, liquid or solid form. The next three chapters are intended to give only the briefest introduction to these fuels, and the reader is referred to the many books giving a more comprehensive treatment of the properties of fuels [1–3], together with the technology involved in their combustion.

There are numerous factors which need to be taken into account when selecting a fuel for any given application. Economics is the overriding consideration – the capital cost of the combustion equipment together with the running costs, which are fuel purchasing and maintenance. It is impossible to state any firm trends – in recent years the price of oil has varied between a bargain and the almost prohibitive, for many heating applications. The position is often complicated locally by price distortions arising from taxation policy.

## 7.2 Natural Gas

Most of the gas burnt in thermal installations consists of natural gas. This is obtained from deposits in sedimentary rock formations which are also sources of oil. Natural gas is extracted from production fields and piped (at approximately 90 bar) to a processing plant where condensable hydrocarbons are extracted from the raw product. It is then distributed in a high-pressure mains system. Pressure losses are made up by intermediate booster stations and the pressure is dropped to around 2500 Pa in governor installations where gas is taken from the mains and enters local distribution networks.

The initial processing, compression and heating at governor installations uses the gas as an energy source. The energy overhead of the winning and distribution of a natural gas is about 6% of the extracted calorific value.

The composition of a natural gas will vary according to where it was extracted from, but the principal constituent is always methane. There are generally small quantities of higher hydrocarbons together with around 1% by volume of inert gas (mostly nitrogen). The characteristics of a typical natural gas are:

|  |  |  |
|---|---|---|
| Composition (%volume) | CH$_4$ (methane) | 92 |
|  | other hydrocarbons | 5 |
|  | inert gases | 3 |

|  |  |
|---|---|
| Density (kg m$^{-3}$) | 0.7 |
| Gross calorific value (MJ m$^{-3}$) | 41 |

# 7.3  Town Gas (Coal Gas)

The original source of the gas which was distributed to towns and cities by supply utilities was from the gasification of coal. In its simplest form the process consisted of burning a suitable grade of coal in a bed with a carefully controlled air supply (and steam injection) to produce gas and also coke, a solid fuel consisting of almost pure carbon, which was used in metallurgical processes. While the energy supply market world-wide is dominated by natural gas, there is still much interest in gas derived from the gasification of coal. This is still the gas supplied by utility companies in many parts of the world (e.g. Hong Kong) and there is continuing longer-term development of coal gasification, since it is one of the most likely ways of exploiting the substantial world reserves of solid fuel.

Coal gas, also known as town gas, was first introduced into the UK and the USA at the beginning of the 19th century. The gas was produced by heating the raw coal in the absence of air to drive off the volatile products. A more effective process which soon followed used partial oxidation of the coal. This was essentially a two-stage process, with the carbon in the coal being initially oxidised to carbon dioxide, followed by a reduction to carbon monoxide:

$$C + O_2 \rightarrow CO_2$$

$$CO_2 + C \rightarrow 2CO$$

The volatile constituents from the coal were also present, hence the gas contained some methane and hydrogen from this source. An improved product was obtained if water was admitted to the reacting mixture, the water being reduced in the so-called *water gas reaction*:

$$C + H_2O \rightarrow CO + H_2$$

This gas was produced by a cyclic process where the reacting bed was alternately blown with air and steam – the former exhibiting an exothermic, and the latter an endothermic, reaction. A typical town gas produced by this process has the following properties:

| Composition (% volume) | $H_2$ | 48 |
|---|---|---|
| | CO | 5 |
| | $CH_4$ | 34 |
| | $CO_2$ | 13 |

| | |
|---|---|
| Density (kg m$^{-3}$) | 0.6 |
| Gross calorific value (MJ m$^{-3}$) | 20.2 |

A more recent gasification process, developed since 1936, is the Lurgi gasifier. In this process the reaction vessel is pressurised, and oxygen (as opposed to air) as well as steam is injected into the hot bed. The products of this stage of the reaction are principally carbon monoxide and hydrogen. Further reaction to methane is promoted by a nickel catalyst at temperatures of about 250–350°C:

$$CO + 3H_2 \rightarrow CH_4 + H_2O$$

If air were to be used as the source of oxygen for the gasification process, significant quantities of nitrogen would be present in the product. The use of oxygen enrichment in the process reduces this, and so increases the calorific value of the product.

The sulphur present in the coal can be removed by the presence of limestone (calcium carbonate) as follows:

$$H_2 + S \rightarrow H_2S$$

$$H_2S + CaCO_3 \rightarrow CaS + H_2O + CO_2$$

## 7.4  Liquefied Petroleum Gas (LPG)

The final type of gaseous fuel that we will consider is LPG, which as its name suggests is a petroleum-derived product distributed and stored as a liquid in pressurised containers. The normal pressure within the container is the saturated vapour pressure of the fuel corresponding to its ambient temperature. The fuel is comparatively expensive, hence its uses tend to be reserved for premium applications such as transport, portable space heating or situations where gas-burning equipment is required and there is no access to a mains supply.

LPG fuels have slightly variable properties, but they are generally based on propane ($C_3H_8$) or the less volatile butane ($C_4H_{10}$). Compared to the gaseous fuel described above, commercial propane and butane have higher calorific values (on a volumetric basis) and higher densities. Both these fuels are heavier than air, which can have a bearing on safety precautions in some circumstances.

Typical properties of industrial liquefied petroleum gases are given below:

| Gas | Propane | Butane |
|---|---|---|
| Density (kg m$^{-3}$) | 1.7–1.9 | 2.3–2.5 |
| Gross calorific value (MJ m$^{-3}$) | 96 | 122 |
| Boiling point (°C at 1 bar) | −45 | 0 |

# 7.5  Combustion of Gaseous Fuels

## 7.5.1  Flammability Limits

Gaseous fuels are capable of being fully mixed (i.e. at a molecular level) with the combustion air. However, not all mixtures of fuel and air are capable of supporting, or propagating, a flame. Imagine that a region of space containing a fuel/air mixture consists of many small discrete (control) volumes. If an ignition source is applied to one of these small volumes, then a flame will propagate throughout the mixture if the energy transfer out of the control volume is sufficient to cause ignition in the adjacent regions. The implication here is that the heat generated in the control volume must exceed the losses from it.

Clearly the temperature generated in the control volume will be greatest if the mixture is stoichiometric, whereas if the mixture goes progressively either fuel-rich or fuel-lean, the temperature will decrease. When the energy transfer from the initial control volume is insufficient to propagate a flame, the mixture will be non-flammable. This simplified picture indicates that there will be upper and lower flammability limits for any gaseous fuel, and that they will be approximately symmetrically distributed about the stoichiometric fuel/air ratio. A comprehensive review of flammability limits has been published by Zabetakis [4].

Flammability limits can be experimentally determined to a high degree of repeatability in an apparatus developed by the US Bureau of Mines [5]. The apparatus (Fig. 7.1) consists of a flame tube with ignition electrodes near to its lower end. Intimate mixing of the gas/air mixture is obtained by recirculating the mixture with a pump. Once this has been achieved, the cover plate is removed and a spark is activated. The mixture is considered flammable if a flame propagates upwards a minimum distance of 750 mm. The limits are affected by temperature and pressure but the values are usually quoted as volume percentages at atmospheric pressure and 25°C. Typical values for some gaseous fuels are:

| Fuel | Lower limit (%) | Upper limit (%) |
|---|---|---|
| Methane | 5 | 15 |
| Propane | 2 | 10 |
| Hydrogen | 4 | 74 |
| Carbon monoxide | 13 | 74 |

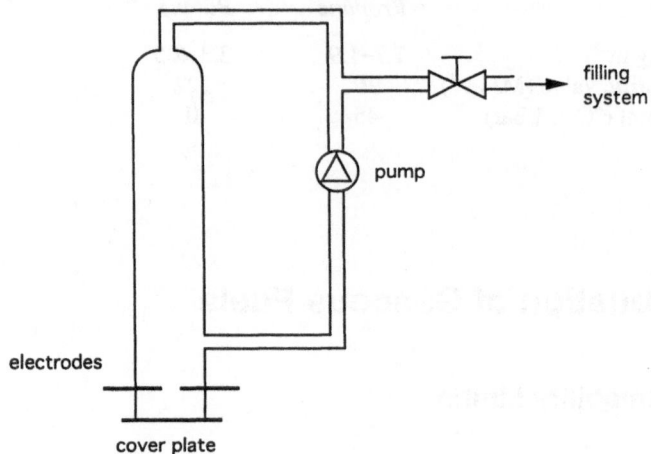

**Figure 7.1** Apparatus for determining flammability limits

The exceptionally wide limits for hydrogen and carbon monoxide are worth noting, as they are significant components of town gas.

## 7.5.2  Burning Velocity

The burning velocity of a gas–air mixture is the rate at which a flat flame front is propagated through its static medium, and it is an important parameter in the design of premixed burners (see below). Strictly speaking, this property should be measured in an unconfined, infinite medium but this is clearly an idealised situation. A simple method of measuring the burning velocity is to establish a flame on the end of a tube similar to that of a laboratory Bunsen burner. When burning in aerated mode, the flame has a distinctive bright blue cone sitting on the end of the tube. The flame front on the gas mixture is travelling inwards normally to the surface of this cone (Fig. 7.2).

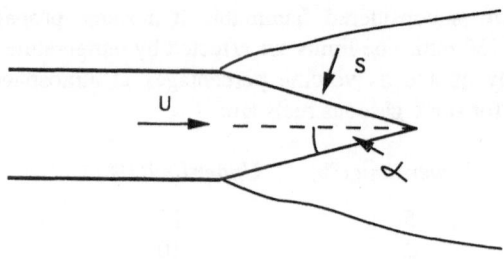

**Figure 7.2** Estimation of burning velocity

If $U$ represents the mean velocity of the gas–air mixture at the end of the tube and $\alpha$ is the half-angle of the cone at the top of the tube, then the burning velocity $S$ can be obtained simply from:

$$S = U \sin(\alpha)$$

The mean velocity $U$ is obtained from the volume flow rate of gas and the cross-sectional area of the end of the tube. This method underestimates the value of $S$ for a number of reasons, including the velocity distribution across the end of the tube and heat losses from the flame to the rim of the tube. More accurate measurements are made with a burner design which produces a flat, laminar flame [6]. Some typical burning velocities are:

| Fuel | Burning velocity (m s$^{-1}$) |
| --- | --- |
| Methane | 0.34 |
| Propane | 0.40 |
| Town gas | 1.0 |
| Hydrogen | 2.52 |
| Carbon monoxide | 0.43 |

Burning velocity should not be confused with the speed of propagation of the flame front relative to a fixed point, which is generally referred to as *flame speed*. This is what would be observed, for instance, if a flame front were being propagated from an ignition source in a room filled with a flammable mixture. In this case, the speed of the flame front is accelerated by the expansion of the hot gas behind the flame.

## 7.5.3 Wobbe Number

The final characteristic of gaseous fuels which will be considered here concerns the interchangeability of one gaseous fuel with another in the same equipment. In very basic terms, a burner can be viewed in terms of the gas being supplied through a restricted orifice into a zone where ignition and combustion take place. The three important variables affecting the performance of this system are the size of the orifice, the pressure across it (or the supply pressure if the combustion zone is at ambient pressure) and the calorific value of the fuel, which determines the heat release rate.

If two gaseous fuels are to be interchangeable, the same supply pressure should produce the same heat release rate. If we consider the restriction to behave like a sharp-edged orifice plate, and if the cross-sectional area of the orifice ($A_o$) is much less than the cross-sectional area of the supply pipe then the mass flow rate of fuel is given by

$$m = C_d A_o (2\rho\Delta p)^{0.5}$$

or in terms of volume flow rate

$$v = C_d A_o \left( \frac{2\Delta p}{\rho} \right)^{0.5}$$

and the heat release rate will be obtained by multiplying this term by the volumetric calorific value of the fuel

$$Q = CV \, C_d \, A_o \left( \frac{2\Delta p}{\rho} \right)^{0.5}$$

If we have two fuels denoted as 1 and 2, we would expect the same heat release from the same orifice and the same pressure drop $\Delta p$, if

$$CV_1 \, C_d \, A_o \left( \frac{2\Delta p}{\rho_1} \right)^{0.5} = CV_2 \, C_d \, A_o \left( \frac{2\Delta p}{\rho_2} \right)^{0.5}$$

i.e.

$$\frac{CV_1}{\rho_1^{0.5}} = \frac{CV_2}{\rho_2^{0.5}}$$

This very simplified analysis serves to highlight the significance of the ratio of the calorific value of a fuel to the square root of its density in defining the performance of the fuel in a particular piece of equipment. This ratio is known as the *Wobbe number* of a gaseous fuel and is defined as:

$$\frac{\text{Gross calorific value (MJ m}^{-3})}{\{\text{Relative density (air} = 1)\}^{0.5}}$$

Some typical Wobbe numbers are:

| *Fuel* | *Wobbe number* (MJ m$^{-3}$) |
| --- | --- |
| Methane | 55 |
| Propane | 78 |
| Natural gas | 50 |
| Town gas | 27 |

The significant difference between the values for natural gas and town gas illustrates why appliance conversions were necessary when the UK changed its mains-distributed fuel in 1966.

## Example 7.1

Calculate the Wobbe number for a by-product gas from an industrial process which has the following composition by volume:

| | |
| --- | --- |
| $H_2$ | 12% |
| $CO$ | 29% |
| $CH_4$ | 3% |
| $N_2$ | 52% |
| $CO_2$ | 4% |

The gross calorific values are:

CO     11.85 MJ m$^{-3}$
CH$_4$   37.07 MJ m$^{-3}$
H$_2$     11.92 MJ m$^{-3}$

First we establish the calorific value of the mixture – this is done on a volumetric basis, so:

$$CV = (0.12 \times 11.92) + (0.29 \times 11.85) + (0.03 \times 37.07) = 5.98 \text{ MJ m}^{-3}$$

The low figure is due to the high proportion of incombustible gases in the mixture.

The relative density of the mixture is calculated by dividing the mean molecular weight of the gas by the corresponding value for air (28.84). The mean molecular weight of this mixture is:

$$(0.12 \times 2) + (0.29 \times 28) + (0.03 \times 16) + (0.52 \times 28) + (0.04 \times 44) = 25.16$$

The relative density is thus $25.16 \div 28.84 = 0.872$. The Wobbe number is then:

$$\frac{5.98}{(0.872)^{0.5}} = 6.36$$

The Wobbe number of a fuel is not the only factor in determining the suitability of a fuel for a particular burner – the burning velocity of a fuel is also important. In general, any device will operate within a triangular performance r ap, such as that sketched in Fig. 7.3. Outside the enclosed region, combustion characteristics will be unsatisfactory in the way indicated on the diagram.

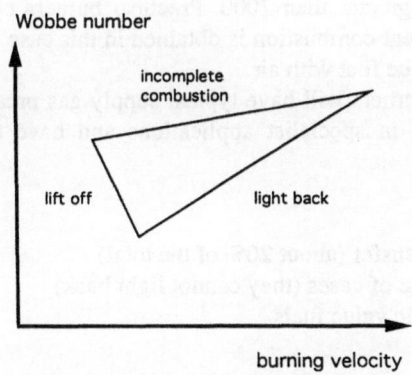

Figure 7.3 Gas burner operation with different fuels

# 7.6 Gas Burners

## 7.6.1 Diffusion Burners

Diffusion burners operate on a very simple principle – the fuel issues from a jet into the surrounding air and the flame burns by diffusion of this air into the gas envelope (Fig. 7.4). Combustion will occur where the local air to fuel ratio is within the flammability limits. This type of flame is commonly observed in a laboratory Bunsen burner when the air hole at the bottom is closed.

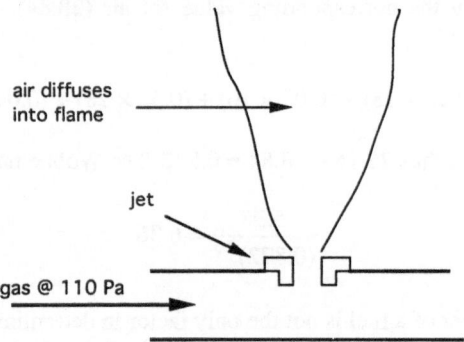

air diffuses
into flame

jet

gas @ 110 Pa

**Figure 7.4** Diffusion burner

A diffusion flame from a hydrocarbon fuel has a yellow colour as a result of radiation from the carbon particles which are formed within the flame. The flame can have laminar characteristics or it may be turbulent if the Reynolds number at the nozzle of the burner is greater than 2000. Practical burners operate in the turbulent regime since more efficient combustion is obtained in this case because the turbulence improves the mixing of the fuel with air.

Industrial diffusion burners will have typical supply gas pressures of 110 Pa. They are normally used only in specialist applications and have the following positive characteristics:

Quiet operation
High radiation heat transfer (about 20% of the total)
Will burn a wide range of gases (they cannot light back)
Useful for low calorific value fuels

## 7.6.2 Premixed Burners

The vast majority of practical gaseous burners mix the air and fuel before they pass through a jet into the combustion zone. In the simplest burners, such as those that are used in domestic cookers and boilers, the buoyancy force generated by the hot gases is used to overcome the resistance of the equipment, but in larger installations the gas supply pressure is boosted and the air is supplied by a fan. The principle is illustrated by the flame from a Bunsen burner with the air hole open, and is shown diagram-

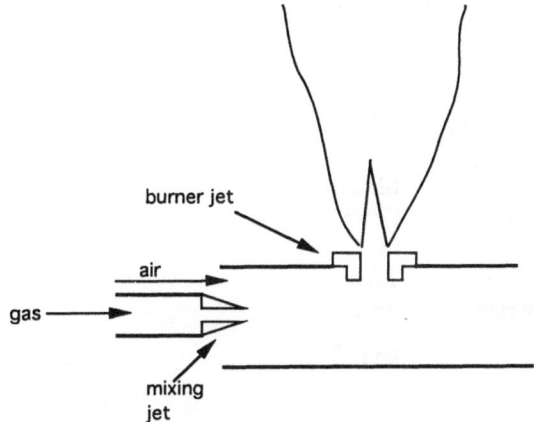

**Figure 7.5** Premixed burner

matically in Fig. 7.5. The gas and air are mixed between the fuel jet and the burner jet, usually with all the air required for complete combustion. The velocity of the mixture through the burner jet is important: if the velocity is too low (below the burning velocity of the mixture) the flame can light back into the mixing region, whereas if the velocity is too high the flame can 'lift off' from the burner to the extent where it can be extinguished by, for instance, entrainment of additional (secondary) air around the burner.

The flame from a premixed burner will emit very little heat by radiation but, because of its turbulent nature, forced convection in a heat exchanger is very effective.

# 7.7 References

1. Samir S (1990). Fuels and Combustion. Orient Longman, London
2. Harker JA (1972). Fuel Science. Oliver and Boyd, London
3. Goodger EM (1975). Hydrocarbon Fuels. Macmillan, London
4. Zabetakis MG (1965). Flammability Characteristics of Combustible Gases and Vapours. US Bureau of Mines Bulletin 627
5. Coward HF, Jones GW (1952). Limits of Flammability of Gases and Vapours. US Bureau of Mines Bulletin 503

6. Botha JP, Spalding DB (1954). The laminar flame speed of propane/air mixtures with heat extraction from the flame. Proc Roy Soc A225: 71–96

## 7.8  List of Symbols

| | | |
|---|---|---|
| $A$ | cross-sectional area | $m^2$ |
| $C$ | coefficient | – |
| CV | calorific value | $MJ\ m^{-3}$ |
| $m$ | mass flow rate | $kg\ s^{-1}$ |
| $p$ | pressure | Pa |
| $Q$ | heat release rate | kW |
| $v$ | volume flow rate | $m^3\ s^{-1}$ |
| $\rho$ | density | $kg\ m^{-3}$ |

*Subscripts*

| | |
|---|---|
| d | discharge |
| o | orifice |
| 1 | initial state |
| 2 | alternative state |

# Chapter 8

# Liquid Fuels

---

## 8.1 Occurrence and Processing

The basic source of liquid fuels is crude oil, which occurs in strata of sedimentary rocks. Gaseous, liquid and semi-solid materials are separated at the well head, but the liquid fuels which are burned in practice are first processed in a refinery.

The most significant of the refinery processes is distillation. The crude oil is flashed (i.e. it undergoes a sudden drop in pressure) into a column. The main part of the column is at atmospheric pressure with a vacuum section producing the heavier products. The vaporised oil travels up the column (Fig. 8.1) which has a vertical temperature gradient with the top of the column being the coolest part. The fraction of the oil which vaporises in the column will condense out at the appropriate level in the column. A system of bubble caps and trays is used to facilitate this.

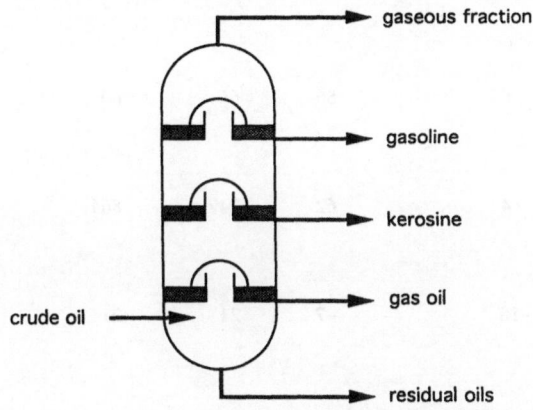

**Figure 8.1** Schematic of distillation column

The most volatile fractions, for example fuels for internal combustion engines, are extracted from the top of the column. All the products which have been condensed in the column are grouped as *distillate* oils, whereas the component of the feedstock which did not evaporate forms *residual* fuel products. The simple distillation of a crude oil is termed a *straight run*, and the relative proportions produced are unlikely to optimise use of the raw product. In order to get a better match to the market requirements, further processing will take place which enables heavier fractions to be 'cracked' into lighter products or the lower molecular weight components reformed into larger molecules.

## 8.2  Properties of Oil Fuels

Both distillate and residual fuel oils are burned in thermal plants; there are various nationally accepted methods for classifying their properties, such as BS2869 in the UK [1]. Table 8.1 gives some physical properties of the commercially important fuel oils: note that classes C and D are distillate oils, whereas E, F and G are all grades of residual oil. Some explanatory notes about the properties in the table follow.

**Table 8.1**  Typical properties of liquid fuels

| | Type | | | | |
|---|---|---|---|---|---|
| | Distillate oils | | Residual fuel oils | | |
| | Kerosine Class C | Gas oil Class D | Light Class E | Medium Class F | Heavy Class G |
| Specific gravity @ 15.5°C | 0.79 | 0.84 | 0.93 | 0.95 | 0.97 |
| Flash point (°C) | 38 | 66 | 66 | 66 | 66 |
| Kinematic viscosity @ 38°C (cS) | 2 | 4 | 62 | 247 | 864 |
| Pour point (°C) | – | −18 | −7 | 21 | 21 |
| Calorific value (MJ $kg^{-1}$ $K^{-1}$) (gross) | 46.4 | 45.5 | 43.3 | 42.9 | 42.5 |
| Sulphur (% by mass) | 0.2 | 0.75 | 2.75 | 3.25 | 3.30 |
| $c_p$ (0–100°C) (kJ $kg^{-1}$) | 2.1 | 2.06 | 1.93 | 1.89 | 1.89 |
| Relative cost (gas oil = 1) | 1.15 | 1.0 | 0.87 | 0.77 | 0.72 |

## Flash Point

This is usually quoted in terms of a standard test (Pensky–Martens [2]) which gives an indication of the flammability of the fuel. Its significance is in the safety aspects of storing and handling the fuel.

## Viscosity

This is a measure of resistance to flow and is therefore very significant as far as boiler installations are concerned; it reflects the energy required to pump the oil through pipework and it has an important bearing on the atomisation process in burners.

The usual method of quoting the viscosity of an oil is to give the value of the kinematic viscosity in centistokes (cS). Kinematic viscosity is the dynamic viscosity divided by the density of the fluid; it is measured in a standard U-tube viscometer at 80°C.

## Pour Point

This is complementary to viscosity in that it gives an indication of the temperature at which the oil will start to flow freely. Note the significant difference between the distillate and residual oils.

## Calorific Value

The calorific value of a liquid fuel is measured in a bomb calorimeter, which measures directly the gross calorific value at constant volume. Although strictly speaking a constant-pressure value is appropriate for boiler applications, the two figures are very close (see Chapter 3). The difference between gross and net values, however, must never be ignored! It can be seen that the less volatile oils have lower calorific values.

## Sulphur Content

Sulphur exists in all liquid fuels, but it is present to a significant degree in residual fuel oils. When burnt, sulphur forms sulphur dioxide ($SO_2$) and sulphur trioxide ($SO_3$) which are major sources of air pollution (see Chapter 10). There are two important consequences of this for thermal plant installations:

1. the flue must be designed to provide acceptable concentrations of $SO_x$ at ground level;
2. it is particularly important to prevent condensation from the flue gases anywhere in the equipment, as both $SO_2$ and $SO_3$ are soluble in water, forming sulphurous and sulphuric acid respectively, and sulphuric acid vapour can be formed in the flue gas.

## Specific Heat

A knowledge of the specific heat of the liquid is important in handling liquid fuels since the residual oils all have to be heated before they can be atomised, and the heavier grades must be stored in a heated tank if they are to flow freely into the distribution pipework.

## Relative Cost

This is a very rough indication of the comparative costs of the oils. These figures change in response to market and fiscal trends but it can be seen that there is a clear price differential in favour of the residual oils, which in part reflects the premium value and flexibility of use of the lighter distillate fuels.

The capital investment in the storage system, handling and combustion equipment for residual fuel oils is significantly higher than that needed for distillate oils, hence an economic optimisation is necessary when making a decision as to the most appropriate fuel for a given application.

# 8.3  Combustion of Liquid Fuels

For efficient combustion, a liquid fuel must be broken up into a stream of droplets to maximise the surface area-to-volume ratio. In small domestic equipment, light distillate oils can be vaporised in a small 'pot' burner, but in anything larger than this the fuel must be atomised. The various types of burner accomplish this in different ways, but the objective of all types of burner is to produce a spray of droplets which are small, and which have a narrow size distribution.

The combustion process consists of evaporation of the droplet, driven by heat transfer from its surroundings, with the vapour subsequently burning in a diffusion flame. Studies of individual burning droplets [3] have highlighted the relationship between heat and mass transfer at the surface of the burning droplet. A classical analysis of this situation where the heat transfer to the droplet is by convection [4] shows that the rate of mass transfer (and hence the combustion rate of the droplet) is inversely proportional to the diameter. This emphasises the importance of producing a fine spray of droplets in the burner. Because evaporation of the droplet is the controlling influence on the combustion rate, liquids with low latent heats of evaporation will burn more quickly. The combustion of a liquid fuel spray is more complex than is indicated above: in particular, heat transfer to the droplet will be by radiation as well as convection.

The combustion of oil is a two-phase process; intimate mixing of the fuel and air is an important requirement hence the fuel is broken up into a fine spray, or *atomised*.

The different types of burners are distinguished by the way in which they atomise the fuel and the most common methods are briefly reviewed here.

## 8.4 Pressure Jet Burners

The simplest form of pressure jet consists of a swirl chamber through which the fuel passes before issuing through the final orifice (Fig. 8.2). Angular velocity is imparted to the liquid by tangential slots or ports. The fuel emerges from the jet as a conical sheet which subsequently breaks up into droplets of between 10 and 200 μm diameter. The oil supply pressure is usually greater than 500 kPa.

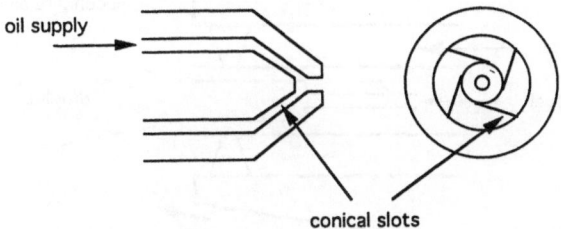

**Figure 8.2** Simplex pressure jet

The simplest and most common type of oil burner incorporates the pressure jet, and this is shown diagrammatically in Fig. 8.3. Pressure jet burners span a wide range of ratings, from domestic units of about 20 kW up to 2.5 MW. Classes C, D and E oils can be burned.

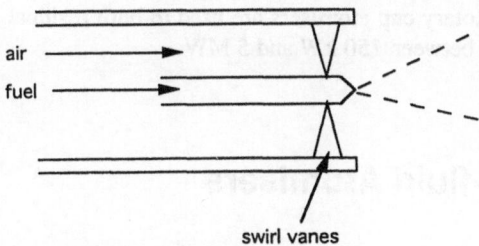

**Figure 8.3** Pressure jet burner

The control of the output of the burner depends on its size. Units of less than 300 kW usually operate in on/off mode; larger units than this can incorporate continuous modulation of the air and oil supply flow rates, with reversion to on/off control at low loads. Pressure jet burners are not capable of modulating to accommodate a wide range of loads. A maximum turndown ratio (max. firing rate:min. firing rate) of 2:1 can be achieved, but 1.5:1 is a more usual figure.

## 8.5  Rotary Cup Burners

In this type of equipment the supply of oil is fed onto a rotating surface (usually a cup or disk) and the atomisation is achieved when the fluid is flung off the cup by centrifugal force (Fig. 8.4). These atomisers tend to give a narrow size range of droplets and are ideally suited to the more viscous liquid fuels, as pumping pressures are much lower than those for pressure jet burners. The cup rotates at 4000–6000 rpm to atomise a class G residual fuel oil, although much higher speeds are used in some applications.

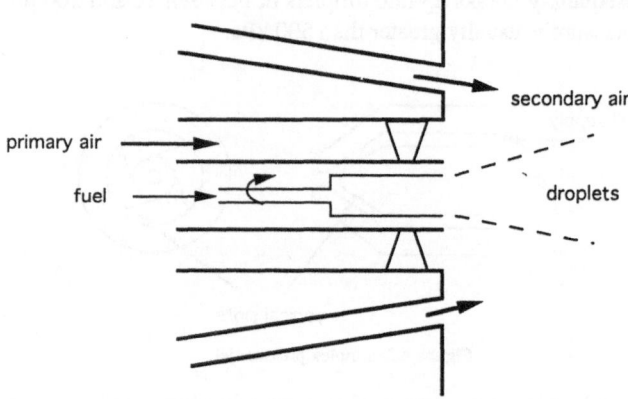

**Figure 8.4** Rotary cup burner

The air supply to this type of burner is split into two streams: 15% is supplied as primary air around the atomiser itself, the remainder being admitted subsequently as secondary air. Rotary cup atomisers are used to burn residual oils of classes E, F and G, and are rated between 150 kW and 5 MW.

## 8.6  Twin-fluid Atomisers

In this type of atomiser a secondary fluid (typically air or steam) is used to produce the shear necessary to break up the oil into droplets. The nozzle is essentially similar to that of the pressure jet burner (Section 8.4) with the addition of an extra set of tangential ports on the inside or outside of the oil flow passage. In an air-blast atomiser about 2–10% of the combustion air is supplied at a high pressure (20–100 kPa). This type of atomiser is more expensive to operate than the other types but it is capable of much greater load modulation, achieving turndown ratios up to 5:1, which can make the extra expense worthwhile.

# 8.7  Storage of Liquid Fuels

Provision must be made for the storage of oil fuels. Tanks may be located inside or outside and are usually of the vertical or horizontal cylindrical type. It is usual to allow for a storage capacity equivalent to 2–3 weeks' full load running – this figure is easily obtained from the rating of the burner and the calorific value of the fuel.

The oil tank itself must be vented, as must the tank room. If the tank is inside, a bund wall must be provided which can hold the entire tank contents. Pipework is usually of mild steel with iron fittings.

For distillate oil systems a single pipe system is adequate for delivery of the oil to the burner. For residual fuel oil systems a ring main system is required and provision must be made to ensure that the oil is at the appropriate temperature, as summarised in Table 8.2. Additional heating is usually provided at the burner appropriate to the type of atomiser in use. Where storage heating is required (classes F and G) the tank is insulated against heat loss.

**Table 8.2**  Storage and handling temperatures for liquid fuels

| Class | Min. temperature (°C) | |
|---|---|---|
| | Storage | Outflow |
| E | 10 | 10 |
| F | 25 | 30 |
| G | 40 | 50 |

# 8.8  References

1. British Standards Institution (1983). BS 2869: Specification for fuel oils for oil engines and burners for non-marine use. BSI, London
2. Institute of Petroleum (1988). Standard Methods for the Analysis and Testing of Petroleum Products. Wiley, Chichester.
3. Williams A (1973). Combustion of droplets of liquid fuels: a review. Combustion and Flame 21: 1–31
4. Spalding DB (1955). Some Fundamentals of Combustion. Butterworth, London

# Chapter 9

# Solid Fuels

## 9.1 Introduction

Solid fuel embraces a wide variety of combustibles, ranging from wood, peat and lignite, through refuse and other low calorific value substances, to coal and other solid fuels derived from it. Coal represents by far the largest component of the world's fossil fuel reserves. In thermal terms 90% of the known hydrocarbon fuel deposits are formed by coal. The carbon:hydrogen ratio of coal is the highest of the fossil fuels, hence the calorific values of coals are principally determined by the carbon in the fuel. For the same reason, coal fuels produce the highest emissions of carbon dioxide.

The properties of coal vary widely and a classification system is used to identify particular fuels. It is usual to consider coals in terms of their *rank*: in general, a high ranking coal will have a high carbon content. The other major coal constituent element, hydrogen, is present in hydrocarbons which are released as *volatile matter* when the coal is heated.

Coal is a sedimentary rock of vegetable origin. Vast deposits of plant material, formed approximately 80 million years ago, were consolidated by pressure, heat and earth movement. The rank of a coal is related to its geological age and, generally, its depth in the earth. The ranking sequence is

Wood
Peat
Lignite (brown coal)
Bituminous coal
Anthracite

with bituminous coal being used in the bulk of heating plant applications. The location of the coal deposits has an important influence on the economic viability of their use. In general, deposits close to the surface which can be worked by strip mining produce a more economical fuel than deep mined coal.

Coal was the fuel which fired the Industrial Revolution, but it is no longer the cheapest option among the fossil fuels. The cost of working the deposits and the investment in technology needed to meet increasingly stringent emissions standards have increased the cost of burning coal. However, the sheer quantity of known coal deposits, combined with an ever-increasing world demand for energy, means that exploitation in one form or another is inevitable. Recent developments in gasification processes have shown that it is possible to produce gas from coal at a viable thermal efficiency and to remove the sulphur from the fuel at the same time; this could prove a way forward to the utilisation of these fuel reserves.

# 9.2  Coal Classification

As the rank of a coal increases, its carbon content increases from 75% to about 93% (by weight), the hydrogen content decreases from 6% to 3%, and the oxygen content decreases from around 20% to 3%.

A useful method for analysing a coal is the *proximate* process. This yields the constitution of the coal in terms of its volatile matter, moisture content and ash content [1]. Proximate analyses of some common fuels are given in Table 9.1. The moisture in coal is made up of two components: surface moisture and inherent moisture. The former is affected by the way in which the coal is stored, and is thus variable.

Table 9.1  Composition of some typical solid fuels (% by mass)

| Fuel | Carbon | Volatile matter | Moisture | Ash |
|------|--------|-----------------|----------|-----|
| Peat | 44 | 65 | 20 | 4 |
| Lignite | 57 | 50 | 15 | 4 |
| Bituminous coal | 82 | 25 | 2 | 5 |
| Anthracite | 90 | 4 | 1 | 3 |

Coals are also analysed in terms of their elemental constituents, giving the *ultimate* analysis which was used earlier in stoichiometric calculations. Typical ultimate analyses of two types of solid fuel are given in Table 9.2.

Table 9.2  Ultimate analyses (% by mass) of some coals [2]

| | Carbon | Hydrogen | Oxygen | Nitrogen | Sulphur |
|------|--------|----------|--------|----------|---------|
| Anthracite | 94.4 | 2.9 | 0.9 | 1.1 | 0.7 |
| Bituminous | 89.3 | 5.0 | 3.4 | 1.5 | 0.8 |

## 9.3  Coal Properties

There are a number of properties which are important in identifying the suitability of a coal for any given application; some of these are discussed briefly here.

### Size

The size classification of coals as delivered is historically related to the Imperial system of units. Some common size groups, together with their rather picturesque names, are given in Table 9.3.

### Calorific Value

The ranking of a coal is not necessarily related to its calorific value. Coal fuels generally have a range of values from 21 to 33 MJ $kg^{-1}$ (gross). The design rating of a coal-fired burner is usually based on an estimated calorific value of 26 MJ $kg^{-1}$.

### Ash Fusion Temperature

The melting point of the ash left after combustion of the coal is of particular importance in terms of the combustion and ash disposal equipment. If the ash fuses it produces a glassy, porous substance known as clinker. The combustion equipment will be designed to handle either clinker or unfused ash, and use of the wrong type of coal can have dire consequences.

**Table 9.3**   Size distributions for coals

| Name | Upper limit (mm) | Lower limit |
|------|------------------|-------------|
| Large cobbles | > 150 | 75 |
| Cobbles | 100–150 | 50–100 |
| Trebles | 63–100 | 38–63 |
| Doubles | 38–63 | 25–38 |
| Singles | 25–38 | 13–18 |

### Sulphur Content

Many deep-mined coals have a fairly high sulphur content, typically around 1.5% by weight. The same considerations apply to coal-fired installations as to oil-fired combustion equipment, namely that condensation inside the plant must be avoided and that the design of the flue must ensure that ground concentrations of sulphur oxides are controlled within acceptable limits.

# 9.4 Coal Combustion

Coal combustion is a two-phase process and the objective of the burner is, as always, to achieve complete combustion of the fuel with maximum energy efficiency. Three ways of burning solid fuels are currently in use: pulverised fuel combustion, fluidised bed combustion and burning on a grate. These are briefly reviewed below.

### Pulverised Fuel

As its name implies, the coal is ground to a very fine size (about 0.08 mm) when it can be made to behave rather like a liquid if air is blown upwards through the powder. The preparation and handling equipment is very expensive and pulverised fuel installations are generally only economically viable in very large scale applications, such as thermal power stations. The fuel is injected in the form of a conical spray, inside a swirling conical primary air supply in a fashion analogous to that for an oil burner.

### Fluidised Bed Combustors

Fluidised bed combustors have been the subject of intensive research and development over the last few years. The basic principle of operation is that the coal is mixed with an inert material (e.g. sand) and the bed is 'fluidised' by an upwards flow of air (Fig. 9.1). Although this requires more fan power than the more conventional grate combustors (see below) there are a number of advantages in fluidised bed combustion:

1. the bed temperature can be kept cooler than in the case of grate combustion – fluidised bed temperatures are generally within the range 750–950°C, and this means that ash fusion does not occur and the low temperatures produce less oxides of nitrogen ($NO_x$);
2. high rates of heat transfer can be attained between the bed and heat exchanger tubes immersed in it;
3. a wide range of fuel types can be burned efficiently;
4. additives (such as limestone) can be used which react with oxides of sulphur retaining the sulphur in the bed with consequent reduction in $SO_x$ emissions.

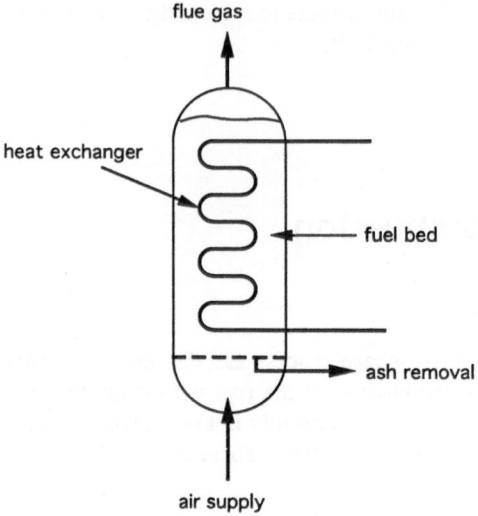

**Figure 9.1** Fluidised bed combustor

## Grate Combustion

The simplest, and most common, way of burning coal is by igniting a bed of the fuel on a porous grate which allows air to rise through the bed, either by buoyancy in smaller equipment or with fan assistance in the larger, automatic stokers.

The combustion of a coal on a grate commences with heat transfer to the raw coal from the adjacent incandescent fuel. The first effect that this has is to drive off the volatile matter from the coal. The volatiles will then rise through the bed, partly reacting with the hot carbonaceous material above it, to burn as a yellow flame above the bed. As the combustion process proceeds, the volatile matter decreases until there is only the carbonaceous residue left, which burns with the intense emission of radiation.

As the air enters the fuel bed from below, the initial reaction is the combustion of the carbon to carbon dioxide. In the hot upper region of the gas this is reduced to carbon monoxide

$$CO_2 + C \rightarrow 2CO$$

which burns in the secondary air above the bed. The effect of this is to decrease the concentration of oxygen from 21% at entry, to zero at about 100 mm above the grate. At this point, there is a peak in the carbon dioxide concentration which falls away as the reduction to carbon monoxide proceeds (Fig. 9.2).

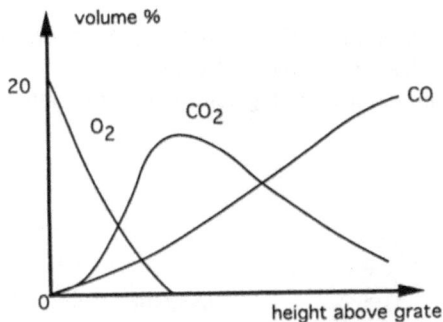

**Figure 9.2** Concentration profiles above a grate

## Underfeed Stoker

The operation of an underfeed stoker is shown in Fig. 9.3. Coal is fed into the retort by the action of a screw. When combustion is completed at the top of the bed, a residue of ash and clinker remains which falls to the sides of the retort.

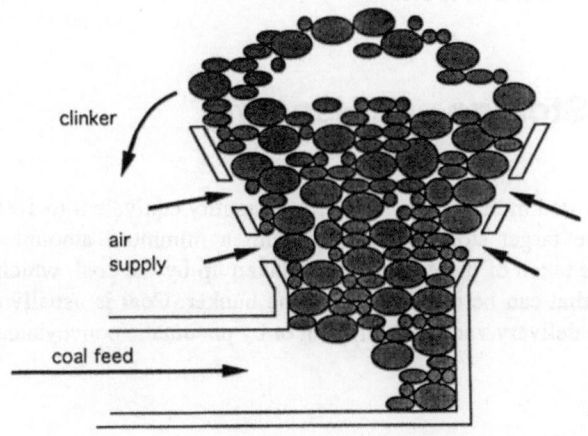

**Figure 9.3** Underfeed stoker

The de-ashing of underfeed stokers is generally a manual process, although some manufacturers offer automatic ash-handling facilities. The ash is raked from a pit underneath the retort and it is particularly important to pay attention to access provision in the installation of this type of burner.

Bituminous singles with an ash fusion temperature of around 1200°C are an appropriate fuel for this type of device.

## Chain Grate Stoker

A diagram of a chain grate boiler is shown in Fig. 9.4. The coal is supplied by the travelling grate and the thickness of the bed controlled by the guillotine door. The speed of the grate and an air damper setting control the firing rate.

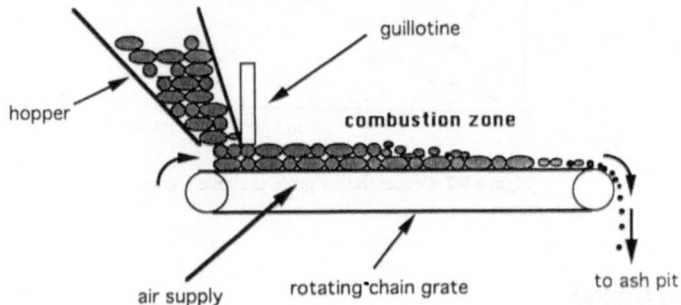

**Figure 9.4** Chain grate stoker

The fuel for such boilers is usually smalls (about 13–25 mm) with a high ash fusion temperature. The ash falls from the end of the grate into a pit, from where it can be removed by a conveyor belt or screw.

# 9.5  Coal Storage and Handling

Solid fuels are stored in bunkers – normally a quantity equivalent to 100 hours at peak firing rate is the target storage capacity, with a minimum amount of 20 tonnes. Account must be taken of the conical shape taken up by the coal, which will limit the amount of coal that can be extracted from the bunker. Coal is usually conveyed into storage from the delivery vehicle by tipping or by pneumatic conveyance along pipes.

# 9.6  References

1.  British Standards Institution (1962). BS 1016: Methods for the analysis and testing of coal and coke. BSI, London
2.  Spiers HM (1962). Technical Data on Fuel. British National Committee, World Power Conference, London

# Chapter 10

# Emissions from Combustion Plant

## 10.1 Environmental Considerations

The environmental effects of the combustion of fossil fuels have assumed a high public profile in recent years. Initially, popular attention has been drawn to the problems through a series of incidents; for example, the London 'smogs' in the earlier part of the twentieth century, and in the latter part the problems, such as photochemical smog, experienced in Los Angeles and the destruction of vegetation by acid rain [1] in northern Europe.

Emissions have come to be seen in a wide context, with scientific credibility being given to concerns about long-term irreversible changes to the climate of the planet – in particular the issue of global warming. The harmful aspects of emissions can be categorised by their perceived consequences: long-term effects on the climate or changes in air quality producing deterioration in human, animal and plant health. Many artefacts such as buildings are being damaged by deteriorating air quality.

This categorisation gives a framework for the study of emissions from combustion systems, but several chemical compounds in flue gases can contribute to both aspects of the problem. Oxides of nitrogen ($NO_x$) have been implicated in both acid rain and photochemical smog, but they are also a highly effective 'greenhouse gas' and as such are of concern in the context of global warming.

## 10.2 The Greenhouse Effect

There are two gases, water vapour and carbon dioxide, naturally present in our atmosphere, that play a significant role in the radiation balance of the Earth. The Earth receives short-wave radiation from the Sun, and emits long-wave (infra-red) radiation out into space. Water vapour and carbon dioxide absorb some wavelengths in the infra-red and hence intercept much of the outgoing radiation leaving the surface of the Earth. This in turn raises the temperature of these molecules in the atmosphere, hence regulating the temperature of the Earth's surface. The effect is analogous to the way in which panes of glass, which also absorb infra-red radiation, provide a warm environment in greenhouses and other glazed buildings.

The mean surface temperature of the Earth is around 15°C. Without the presence of the 'greenhouse gases' this would be nearer −19°C and the Earth would be a frozen, barren waste. The current concern stems from the effects of the temperature rise which will result from a steady increase in the concentrations of these gases. The principal contributor to this problem is seen as carbon dioxide, but other gases such as oxides of nitrogen, methane and chlorofluorocarbons (CFCs) are much stronger absorbers of infra-red than carbon dioxide and may have an increasingly significant effect on the overall picture.

## 10.3  Carbon Dioxide Emissions

Carbon dioxide is a principal product of the combustion of hydrocarbon fuels. While it would seem a comparatively straightforward matter to calculate the rise in carbon dioxide emissions resulting from the combustion of increasing amounts of fossil fuels, assessing the likely change of atmospheric carbon dioxide concentration is not at all simple. Combustion products are discharged into the atmosphere, but two other significant reservoirs of carbon dioxide exist: the oceans and the biosphere (particularly plant life). The dynamics of this problem are well outside the objectives of this book, and for further information the reader will have to consult other sources, such as Liss and Crane [2].

The present (1990) concentration of carbon dioxide in the atmosphere is around 350 parts per million (ppm) by volume, having increased from around 280 ppm in 1765. On present trends, with $CO_2$ concentration rising by about 0.5% per annum, the 1990 value would be approximately doubled by the year 2130. However, if the contributions to the greenhouse effect by other gases are expressed relative to that of $CO_2$ then we are likely to see an effective doubling of $CO_2$ concentration by the year 2060. The increase in atmospheric $CO_2$ concentration is shown in Fig. 10.1, which is taken from measurements made at Mauna Loa, Hawaii, since 1958 [3].

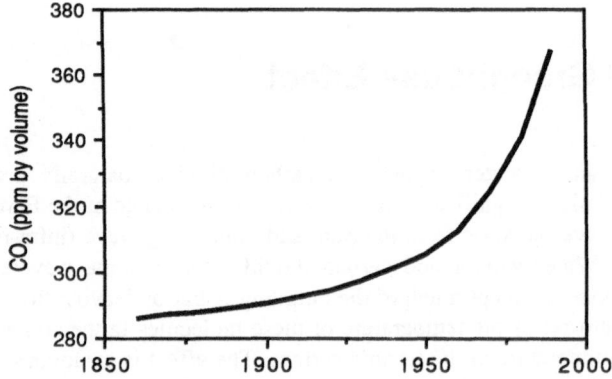

**Figure 10.1**  Carbon dioxide concentration in the atmosphere (after [3])

Clearly the consequences of this, in terms of climatic change, are contentious but the current best estimates, based on coupled ocean–atmospheric models, are that this could result in a rise of 2.5°C in surface temperature. There is much contemporary political debate on appropriate measures to counter the effects predicted by these models. This is a problematical area which is well outside the scope of a technical text such as this, but it is perhaps worth noting that the most obvious course of action is to constrain the growth in the combustion of fossil fuels. Bearing in mind the relationship which exists between economic wealth and energy consumption, there is no sign at present of the economic aspirations of developing countries being counterbalanced by a steady reduction in fuel combustion in the industrialised economies. .

The opportunities for 'technical fixes' in the design and operation of combustion equipment leading to reduced $CO_2$ emissions are very limited. The most obvious course of action, for any given level of energy demand, is to burn the fuel at the highest possible efficiency. This is an economic, rather than a technical, issue and as such is clearly sensitive to legislative pressure. The overall utilisation efficiency of the energy of combustion of fossil fuels can be improved, for example, by the increased use of combined heat and power schemes. A realistic starting point which is emerging in many countries is to stabilise emissions of $CO_2$ at the 1990 level.

A second option which has become the subject of active discussion is to consider a switch, where possible, to fuels with a higher proportion of hydrogen – in particular, natural gas. This will, of course, put pressure on the reserves of this fuel and there is not surprisingly some concern about the use of such a 'premium' fuel for purposes such as the straight generation of electricity.

A comparative performance (in terms of emissions of $CO_2$) is easily derived knowing the percentage of carbon present in the fuel and its calorific value. Natural gas, which has the highest proportion of hydrogen of the common fossil fuels, does emit considerably less carbon dioxide (for a comparable heat release) than other fuels. Comparative $CO_2$ emissions are illustrated in Table 10.1.

**Table 10.1** Comparative $CO_2$ emissions

| Fuel | %C (by mass) | Gross CV (MJ kg$^{-1}$) | kg $CO_2$ (GJ$^{-1}$) (gross) | (net) |
|---|---|---|---|---|
| Natural gas | 0.723 | 54.0 | 49.1 | 49.6 |
| Propane | 0.818 | 50.5 | 59.4 | 60.0 |
| Gas oil | 0.857 | 45.5 | 69.0 | 69.5 |
| Heavy fuel oil | 0.851 | 42.5 | 73.4 | 73.9 |
| Bituminous coal | 0.810 | 33.3 | 89.2 | 89.5 |
| Anthracite | 0.847 | 32.1 | 96.7 | 97.1 |

These figures give some guide as to the comparative emissions of $CO_2$ from different fuels, but it should be borne in mind that variations in the thermal efficiency of plant should also be taken into account.

## 10.4  Carbon Monoxide

All combustion systems emit some carbon monoxide, and its significance is related to its highly toxic nature. The formation of carbon monoxide as a combustion product is generally associated with sub-stoichiometric (i.e. fuel-rich) combustion. In this context carbon monoxide is a particular problem with automobile engines, where it may be present to the extent of several per cent by volume.

Carbon monoxide is a component of the flue gas resulting from hydrocarbon combustion where it exists, given time, in equilibrium with oxygen, hydrogen and water vapour (the water-gas equilibrium). The simplest reacting system is the dissociation equilibrium

$$CO + \tfrac{1}{2}O_2 \rightleftharpoons CO_2$$

discussed in Chapter 5. Knowledge of the dissociation constant for this reaction enables the equilibrium concentration of CO to be estimated, but this is somewhat misleading as higher than equilibrium concentrations are generally measured in boilers and furnaces since the rate of formation of CO is considerably faster than that of its further oxidation to $CO_2$.

In very large systems, equilibrium may be approached, but in smaller, commercial units the concentration of CO in the flue gas is generally about 5–20 ppm by volume. The presence of carbon monoxide is favoured by high temperatures, and inhibited by increased oxygen concentrations.

## 10.5  Oxides of Nitrogen

Oxides of nitrogen have been identified as a significant contributor to air pollution. The convention of expressing them as a chemical 'formula' $NO_x$ in itself suggests that there is a degree of uncertainty as to the exact nature of the species responsible.

The oxides of nitrogen are:

$N_2O$   nitrous oxide
NO    nitric oxide
$NO_2$   nitrogen dioxide

Of these, the two compounds that are of interest in combustion systems are NO and $NO_2$. Nitrogen dioxide in the atmosphere is mostly formed by oxidation of the NO which is discharged in combustion products. Automobile engines have been identified as a prime source of $NO_x$ which is implicated in the formation of photochemical smog, but the enforcement of emission standards for cars means that increasing attention is being paid to the emissions of $NO_x$ from stationary sources.

The greatest proportion of $NO_x$ in combustion systems has been found to be NO, and hence a considerable amount of experimental and theoretical work has been directed at understanding the mechanism of the formation of NO in flames, and the prediction of NO concentrations in combustion systems. There are two routes by which $NO_x$ is formed in flames. The first is via the oxidation of atmospheric nitrogen in the combustion zone. The formation of this NO is very temperature dependent and it is thus often referred to as 'thermal' NO. The second mechanism for the formation of NO is from the nitrogen which is chemically bound within the fuel (nitrogen is generally found in solid fuels as compounds such as pyridine and amines). The $NO_x$ resulting from fuel fragments containing nitrogen is analogously known as 'chemical' $NO_x$.

## 10.5.1 Atmospheric Nitrogen

The oxidation of nitrogen in the combustion air can be represented by the overall balance:

$$\tfrac{1}{2}N_2 + \tfrac{1}{2}O_2 \rightleftharpoons NO$$

The equilibrium constant for this reaction is

$$K_{10.1} = \frac{p_{NO}}{(p_{N_2})^{0.5} (p_{O_2})^{0.5}} \tag{10.1}$$

Given that

$$p_{NO} = (x_{NO}) \cdot p_T$$
$$p_{N_2} = (x_{N_2}) \cdot p_T$$
$$p_{O_2} = (x_{O_2}) \cdot p_T$$

the equilibrium constant simplifies to

$$K_{10.1} = \frac{x_{NO}}{(x_{N_2})^{0.5} (x_{O_2})^{0.5}} \tag{10.2}$$

Knowing the values of $K_{NO}$ as a function of temperature enables the equilibrium concentration of NO to be calculated. The equilibrium concentration of NO can be easily estimated for the case where the fuel is burned with excess air. Under these conditions the concentrations of oxygen and nitrogen in the gas mixture are large compared with the concentration of NO, so they can be assumed to be constant at the values given by a simple stoichiometric calculation. The concentration of nitric oxide is then given by:

$$x_{NO} = K_{10.1}(x_{N_2})^{0.5}(x_{O_2})^{0.5} \tag{10.3}$$

For example, at 1200 K the value of $K_{10.1}$ is 0.005 26. Substituting the stoichiometric values for the concentration of oxygen and nitrogen in the flue gas from the combustion of a natural gas into equation (10.3) gives the following results:

| Excess Air (%) | NO Concentration (ppm) |
|:---:|:---:|
| 5 | 42.5 |
| 10 | 58.9 |
| 15 | 70.9 |
| 20 | 80.4 |
| 25 | 88.5 |
| 30 | 95.3 |

Calculated NO equilibrium concentrations are of limited value because the reactions concerned are slow, hence equilibrium will only be reached when the residence time is long, for example in a very large boiler.

The actual mechanism whereby atmospheric nitrogen is oxidised has been the subject of a great deal of research. It does not involve the simple interaction between molecular nitrogen and oxygen suggested by the equilibrium relationship. A simple chain reaction for the formation of NO was proposed by Zeldovich [4] and this mechanism is thought to be the main route for the formation of thermal NO in combustion systems. The chain reaction is initiated by oxygen atoms, possibly formed by the dissociation of molecular oxygen:

$$O + N_2 \rightleftharpoons NO + N$$
$$N + O_2 \rightleftharpoons NO + O$$

The formation of NO is influenced by the concentration of oxygen in the system and also by temperature. The nature of the above reaction is such that there is a strong dependence on temperature, and a lesser sensitivity to the oxygen concentration. This points to reducing excess air and flame temperature as practical means of controlling the formation of thermal NO although these two measures are, to some extent, mutually incompatible.

## 10.5.2 Fuel-bound Nitrogen

$NO_x$ emissions from systems burning fuels containing nitrogenous compounds are much higher than those obtained from pure hydrocarbon fuels. The nitrogen-bearing compounds in the fuel are generally of a cyclic structure such as pyridines, and these nitrogen bonds are more easily broken to give reactive sites than are the N—N bonds in atmospheric nitrogen.

The formation of $NO_x$ from fuel-bound nitrogen results from low molecular weight fragments such as $NH_3$, $NH_2$ and HCN. Experimental studies have indicated that the formation of $NO_x$ from these sources is rapid, in fact comparable in speed to the combustion of the fuel itself. This means that it is not possible to control the formation of chemical $NO_x$ by reduction in temperature, and NO concentrations in the flames of nitrogen-bearing fuels significantly exceed the calculated equilibrium values. The NO decays towards the equilibrium state after it leaves the vicinity of the combustion zone.

It has been found that this decay process is quicker for fuel-rich flames and that it is slowed by excess air. This gives rise to a difficult situation as the emission of unburned hydrocarbons and carbon monoxide, together with the desire to maximise the thermodynamic efficiency of the system, favour the operation of combustion systems with excess air.

The nitrogen in the fuel can form either $N_2$ or $NO_x$ in the combustion products. The relative proportion of these depends very much on the amount of nitrogen present in the fuel. If this is low, about 0.1% by weight, then the conversion of nitrogen in the fuel to $NO_x$ is high, whereas for fuels containing higher amounts of nitrogen (> 0.5%) possibly only 50% of the nitrogen forms $NO_x$. Gaseous fuels such as natural gas typically contain less than 0.1% nitrogen, liquid fuels can have up to 0.5% nitrogen content, while the nitrogen content of coals is generally within the range 1–2%.

As the nitrogen content of the fuel increases, the chemical $NO_x$ becomes a more significant contributor to the total emissions, with a corresponding decline in the significance of thermal $NO_x$. This trend is illustrated in Fig. 10.2.

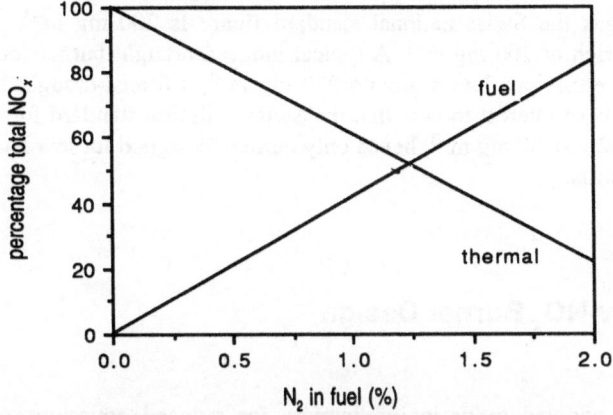

**Figure 10.2** Chemical and thermal $NO_x$ production

There is a growing interest in the emissions of nitrous oxide ($N_2O$) from flames. Although measurements have shown concentrations to be low (within the range 1–10 ppm) from conventional burners, recent research has indicated that higher levels of $N_2O$ (40–250 ppm) may occur in fluidised bed combustion [5]. Nitrous oxide is a particularly effective greenhouse gas and it has also been found to contribute to the depletion of stratospheric ozone.

## 10.5.3 $NO_x$ Levels

The units used for quantifying emissions of oxides of nitrogen have not yet been standardised, but milligrams per normal cubic metre (mg m$^{-3}$) is becoming the most commonly used unit in the EC, although a more familiar unit is parts per million by volume (ppm). Because of the somewhat undefined chemical nature of $NO_x$, the interconversion between the two units is imprecise, but 1 ppm is equivalent to approximately 2 mg m$^{-3}$ (a figure of 2.05 is standard [6]).

EC directive 88/609/EC [7] sets the following emission limits for large plant (> 50 MW):

| Fuel | NO$_x$ Emissions (mg m$^{-3}$) |
|---|---|
| Solid fuels | 650 |
| High-ranking coal | 1300 |
| Liquid fuels | 450 |
| Gaseous fuels | 350 |

The 'tightest' legislation controls currently in force in Europe are the TA-Luft (Germany) and LRV (Switzerland) statutory limits. These laws specify a maximum $NO_x$ concentration of 250 mg m$^{-3}$ for distillate oil-fired burners, although provision is made for further reducing this value if local conditions are deemed to warrant it. For instance, the current standard for Zurich is 120 mg m$^{-3}$. To put this in perspective, the emissions from a traditional pressure jet burner are around 180 mg m$^{-3}$.

For natural gas the Swiss national standard figure is 200 mg m$^{-3}$, with a local standard for Zurich of 100 mg m$^{-3}$. A typical induced-draught burner (domestic type equipment) has emissions levels around 220 mg m$^{-3}$, a forced-draught burner about 105 mg m$^{-3}$. It is of interest to note that the Swiss emission standard for new burners for the year 1992 was 80 mg m$^{-3}$, hence only burners designed for low $NO_x$ emissions met these standards.

## 10.5.4   Low-NO$_x$ Burner Design

The principles adopted in designing burners for reduced emissions of oxides of nitrogen follow from the considerations above – clearly oxygen control and limitation of peak temperatures play a fundamental part. Specific emission control measures currently in use can be summarised:

Staged fuel and/or combustion air
Internal and external recirculation

The three most important factors in limiting $NO_x$ emissions are temperature, oxygen concentration and residence time. A traditional design of burner incorporates single-point injection of the fuel, followed by rapid mixing of the fuel and air. This is often achieved with a swirling air supply. Low $NO_x$ burners split the fuel, the air, or both into stages. A staged fuel/air supply will give a more gradual mixing of the fuel and combustion air, and hence lower peak flame temperatures. The contrast between single-point injection and staged air supply is shown schematically in Fig. 10.3.

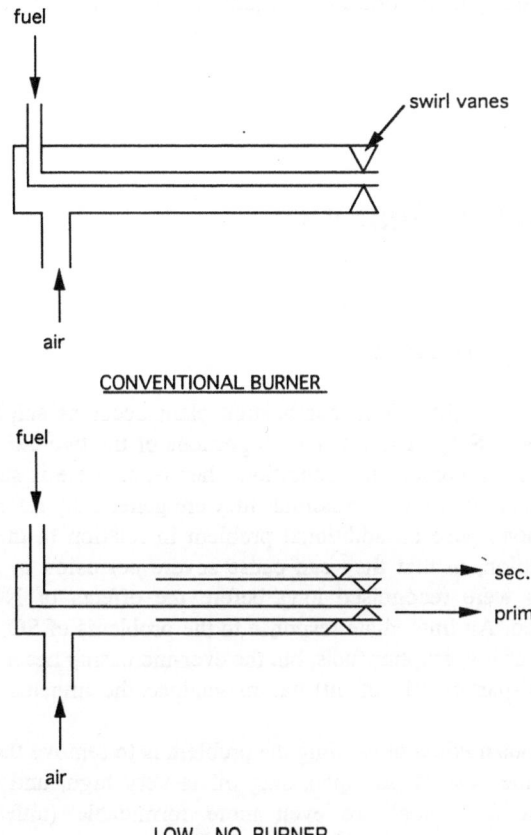

fuel

swirl vanes

air

CONVENTIONAL BURNER

fuel

sec.

prim.

air

LOW – NO$_x$ BURNER

**Figure 10.3** Conventional and low-NO$_x$ burners

In the case of nitrogen-bearing fuels, staged combustion is particularly effective in that, if the nitrogen is liberated in a reducing atmosphere, its oxidation is inhibited and it tends to form molecular nitrogen. The reducing nature of the fuel-rich zones also promotes reduction of any NO$_x$ which may have formed.

Recirculation, either internal, where the gases post flame-front are recirculated to the burner head, or external, where combustion products are returned by a fan from the entry to the flue, also controls the temperature, mixing history and oxygen concentrations in the combustion zone.

Several additional burner configurations are under development which produce a combustion environment favourable to the conversion of fuel nitrogen to N$_2$. These include oil/water atomisation, overfire air and the injection of a material such as ammonia or urea which has a reducing effect on the nitrogen oxides.

There is adequate technology currently available to meet ever-tightening emission standards, but emission control methods are inextricably linked to reduction in combustion intensity. The consequence of this is some derating of boiler equipment and hence future designs of combustion equipment will need to be physically larger than at present.

# 10.6  Oxides of Sulphur

## 10.6.1 The SO$_x$ Problem

Emissions of sulphur oxides from combustion plant occur as sulphur dioxide (SO$_2$) and sulphur trioxide (SO$_3$). The relative proportions of the two oxides are very much dependent on the local combustion conditions; hence, as there is some uncertainty as to the exact formulation of these emissions, they are generically referred to as SO$_x$.

Sulphur emissions pose an additional problem in relation to their contribution to atmospheric pollution, in that they can cause severe corrosion in combustion plant. These difficulties were recognised long before the effects of NO$_x$ on air quality became established. An immediate response to the problems of SO$_x$ formation was to maximise the use of low-sulphur fuels, but the ever-increasing necessity to work lower 'grade' resources (particularly of oil) has maintained the impetus of work on these emissions.

The most obvious method of tackling the problem is to remove the sulphur from the fuel. At present the cost of desulphurising oil is very high, and the difficulties of removing sulphur from coal are even more formidable (unless gasification is undertaken). The removal of SO$_2$ from the flue gas is gradually appearing in large-scale installations (such as power stations) in the form of an absorption process using limestone. This is also not without its environmental complications, as the process increases quarrying activity.

## 10.6.2   Sulphur Trioxide Formation

The formation of sulphur trioxide in combustion systems should be avoided, as it can react with water to form sulphuric acid (H$_2$SO$_4$). The dew point of this acid is generally considerable higher than the water vapour dew point, typically about 150°C, hence any surfaces presented to the combustion gases below this temperature will experience severe corrosion. There is also the possibility of deposition of sulphates, which in turn can cause corrosion and a reduction of heat transfer rates due to fouling.

The SO$_2$/SO$_3$ equilibrium is described by:

$$SO_2 + \tfrac{1}{2}O_2 \rightleftharpoons SO_3 \tag{10.4}$$

The equilibrium constant for this is:

$$K_{10.5} = \frac{p_{SO_3}}{p_{SO_2} (p_{O_2})^{0.5}}$$

(10.5)

which, expressed in terms of mole fractions becomes:

$$K_{10.5} = \frac{x_{SO_3}}{x_{SO_2} (x_{O_2})^{0.5}} \cdot p_T^{-0.5}$$

This can be rearranged as:

$$\frac{x_{SO_3}}{x_{SO_2}} = K_{10.5} (x_{O_2} p_T)^{0.5}$$

$$\frac{x_{SO_3}}{x_{SO_2} + x_{SO_3}} = \frac{K_{10.5} (x_{O_2} p_T)^{0.5}}{1 + K_{10.5} (x_{O_2} p_T)^{0.5}}$$

(10.6)

The formation of $SO_3$ is favoured by higher oxygen concentrations, and equation 10.6 can be solved explicitly to give the proportion of total $SO_x$ as $SO_3$ if the oxygen concentration (governed by the excess air level) is high enough for the oxygen concentration to be approximated as constant. The total number of moles of $SO_x$ from the combustion of unit quantity of fuel is easily obtained from the composition of the fuel, hence the actual equilibrium values for the concentrations of $SO_2$ and $SO_3$ can be easily estimated under excess air conditions.

The equilibrium constant for this reaction is approximated by

$$\log_{10} K_{10.5} = \frac{5014}{T} - 4.755$$

where the partial pressures are expressed in bar [8]. The equilibrium proportions of $SO_2$ and $SO_3$, calculated from equation 10.6, in the flue gas of a boiler burning an oil

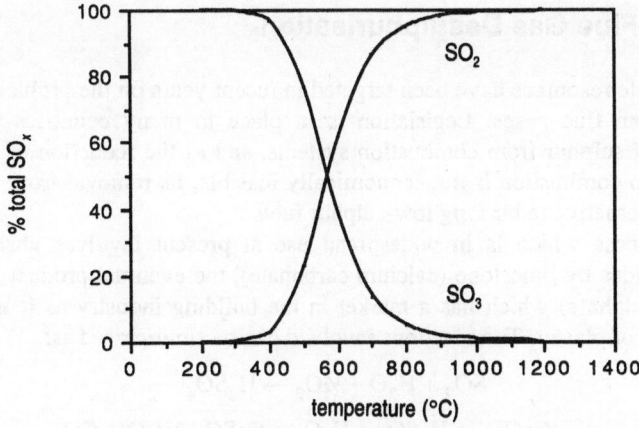

Figure 10.4 Equilibrium $SO_x$ concentrations

containing 2.5% sulphur by weight and operating at 2% excess air is shown in Fig. 10.4.

Residence times in many boilers and other plant are often long enough for equilibrium concentrations to be approached. However, super-equilibrium concentrations of $SO_3$ are frequently measured. The oxidation of $SO_2$ to $SO_3$ can be catalysed by transition metals, such as iron which is used in the construction of the plant, and vanadium which is a low-level constituent of many fuel oils. The actual mechanism by which $SO_3$ is formed is thought to involve atomic oxygen and an inert third body I [9]:

$$SO_2 + O + I \rightarrow SO_3 + I$$

The $SO_3$ that is formed in the gas phase combines with the water vapour in the flue gas to form sulphuric acid

$$SO_3 + H_2O \rightleftharpoons H_2SO_4$$

and at the lower temperatures (about 200°C) experienced in heat exchangers and flues the above equilibrium is almost all sulphuric acid. Clearly the condensation of sulphuric acid onto any surface of the combustion equipment will have very serious consequences, hence it is important to check that sub-dew point temperatures do not occur anywhere, but most especially in the flue system.

The dew point temperature of sulphuric acid is a function of the partial pressure of $H_2SO_4$ in the gas mixture and also depends on the partial pressure of water vapour. It is therefore quite sensitive to the composition of the fuel itself, and empirical expressions for evaluating the acid dew point have been published. A widely accepted expression given by Banchero and Verhoff [10] is

$$\frac{1}{T} = 2.276 \times 10^{-3} - 2.943 \times 10^{-5} \ln p_{H_2O} - 8.58 \times 10^{-5} \ln p_{H_2SO_4} + 6.26 \times 10^{-6} \ln p_{H_2O} \ln p_{H_2SO_4}$$

where $T$ is the acid dew point temperature in degrees Kelvin and the partial pressures are expressed in mm Hg. The dew points predicted by this expression are generally around 200°C.

## 10.6.3   Flue Gas Desulphurisation

Considerable resources have been targeted in recent years on the problem of removing sulphur from flue gases. Legislation is in place in many countries to control the emission of sulphur from combustion systems, and as the reduction of sulphur in the fuel prior to combustion is not economically feasible, its removal from the flue gas is the only alternative to burning low-sulphur fuels.

The process which is in widespread use at present involves absorption of the sulphur oxides by limestone (calcium carbonate), the eventual product being gypsum (calcium sulphate) which has a market in the building industry as it is the principal constituent of plaster. The reactions involved can be summarised as:

$$SO_2 + H_2O + \tfrac{1}{2}O_2 \rightarrow H_2SO_4$$

$$CaCO_3 + H_2SO_4 + H_2O \rightarrow CaSO_4.2H_2O + CO_2$$

The efficiency of removal of sulphur from large installations is usually about 95%.

# 10.7   Particulate Emissions

## 10.7.1   Solid Formation in Combustion

Particulate emissions from combustion systems were one of the first pollutants to attract attention and legislation for their control. The visibility of the discharge of smoke from chimneys, its obvious relationship with air quality and the effects of its deposition on the surrounding environment made smoke an obvious target for environmental improvement.

The particulate matter formed in combustion systems is known by a variety of names, usually relevant to the origin of the material. Smoke, soot, coke, unburned fuel, cenospheres, stack solids and fly ash are all varieties of solid emissions. Hybrid products are also formed, notably the combination of carbonaceous solids with condensable water and sulphuric acid, which can result in the emission of the highly damaging acid smuts from chimneys.

## 10.7.2   Soot Formation

The carbonaceous material known as soot can be formed in the combustion of any hydrocarbon fuel, be it in solid, liquid or gaseous form. The substance is generally familiar as a black powder and its properties are consistent irrespective of its origin, although soot is generally understood to be formed from gaseous combustion.

The presence of soot within a flame can be beneficial in terms of heat transfer; soot has a very high emissivity (approaching that of a black body) and this can much improve the rate of heat transfer in a heat exchanger compared with that which is obtainable with a low-emissivity gas flame. An important consideration is, of course, that any emission of soot from the combustion/heat transfer system represents an energy loss in terms of unburned carbon, hence every effort must be made to enable the combustion of carbonaceous particles to be completed.

The degree to which soot is formed in a combustion system is related to the type of flame and the conditions under which combustion is taking place. Soot formation has been the study of extensive research and a large body of published information exists [11, 12].

The ease with which soot is formed in flames is affected by the nature of the hydrocarbon fuel. Aromatic (ring) compounds have a greater tendency to form soot in flames than, for instance, paraffin hydrocarbons. For this reason the specification of some liquid fuels limits their aromatic content.

Soot that is first formed in flames contains up to about 8% by weight of hydrogen, although this decreases to near 1% later in the combustion process. This corresponds to an empirical formula of $C_8H$, the carbon particles formed agglomerating to form a chain-like structure. Soot is formed under fuel-rich conditions, hence it can be a problem where sub-stoichiometric combustion occurs, such as in the primary combustion zone of an aircraft gas turbine engine.

Measurements of smoke in combustion equipment are made by a number of optical and gas-sampling test procedures, some of which are essentially arbitrary in nature, the best known of these being the Ringelmann test. Contemporary combustion analysers generally have a smoke-testing facility – this operates by drawing a fixed volume of flue gas through a clean filter paper and comparing the darkened disk from the test with a graded template scale.

### 10.7.3   Cenosphere Formation

In the combustion of residual fuel oils there is a significant degree of pyrolysis of the fuel as combustion of the droplets proceeds. If the droplets experience regions of high temperature and sub-stoichiometric conditions then small hollow carbon spheres, known as cenospheres, can be formed. These in turn should burn out in the remainder of the flame, but if this does not happen then unburned carbon can be discharged from the flue. This can amount to 2% by weight of the original fuel and represents a significant thermodynamic loss as well as an environmentally undesirable emission.

### 10.7.4   Acid Smut Formation

A particular problem can occur through the interaction of carbonaceous particle formation and the oxidation of the sulphur compounds present in a fuel. If the conditions are right then carbon particles can agglomerate where the temperature is below the acid dew point (this usually occurs on an internal surface). Although these deposits themselves have a high potential for causing corrosion, after build up they may flake off the attached surface and be emitted via the flue system. Acid smut emissions are highly corrosive and dangerous.

As acid smuts are a combination of sulphur oxides and carbonaceous particles, their control is best approached by addressing the formation of their two constituents: $SO_x$ and carbon, as discussed previously.

### 10.7.5   Ash

Ash is an inorganic material released in the combustion of solid fuels and residual fuel oils. It can occur in molten form or as a dry powder, and, in the former case, can cause corrosion problems particularly in high-temperature zones such as super-heater tubes.

Ash contains silica, together with various metallic compounds of iron, nickel, magnesium and vanadium. In particulate form it can be discharged from chimneys, and where this is a problem, usually in large solid fuel-fired installations, some means of removing it from the flue gas is necessary, such as cyclone separators or electrostatic filters. These can be highly effective, with collection efficiencies of over 90%.

# 10.8  List of Symbols

| | | |
|---|---|---|
| $K$ | equilibrium constant | – |
| $p$ | partial pressure | atmospheres; mm Hg |
| $t$ | temperature | degrees Celsius |
| $T$ | temperature | degrees Kelvin |
| $x$ | mole fraction | – |

# 10.9  References

1. Fisher BEA (1984). Acid rain and where it comes from. Journal of the Institute of Energy 42: 416–20
2. Liss PS, Crane AS (1983). Man-made Carbon Dioxide and Climate Change. Elsevier, Amsterdam.
3. Machta L (1979). Atmospheric measurements of carbon dioxide. Workshop on the Global Effects of Carbon Dioxide from Fossil Fuels, US Department of Energy CONF-770385, UC-11, Washington
4. Zeldovich YB (1946). Acta Physicochem USSR 21: 557
5. Leckner B, Gustavsson L (1991). Reduction of $N_2O$ by gas injection in CFB boilers. Journal of the Institute of Energy 44: 176–182
6. Allen J (1990). Low-$NO_x$ burner systems. Energy World, November: 13–15
7. EC (1988) Large Combustion Plants Directive. EC 88/609, O.J.L. 336
8. Williams A (1990). Combustion of Liquid Fuel Sprays. Butterworth, London
9. Barret RE, Hummel JD, Reid WT (1966). Trans. ASME Ser A 88:165–73
10. Banchero JT, Verhoff FH (1971). J. Inst Fuel 34: 76
11. Palmer HB, Cullis HF (1965). The Chemistry and Physics of Carbon. Marcel Decker, New York
12. Calcote HF (1981). Mechanisms of soot nucleation in flames – a critical review. Combustion and Flame 42: 215–42

# Chapter 11

# Flues and Chimneys

## 11.1 Functions of the Flue System

The function of the flue in combustion equipment can be summarised quite simply: it is for the safe and effective disposal of the products of combustion. In turn this focuses on two considerations, namely bringing the products of combustion to the outlet of the flue at the required conditions (such as temperature and velocity), and ensuring that the location of this outlet is such that the environmental impact of the discharge is controlled. The flue must therefore be regarded as an integral part of the combustion system, and also as a component of its local environment.

There are two ways in which the dispersal of the combustion products can be effected. If the fuel has a very low sulphur content (such as natural gas) then it is often possible to dilute the products of combustion with ambient air and to discharge the diluted mixture at a low level. This may be desirable economically as it avoids the need to construct a high-level discharge, or it might be aesthetically appealing if the presence of a chimney, either outside the building or appearing at the rooftop, is thought to be undesirable.

The majority of flue systems discharge the combustion products at a high level via a flue or chimney. Contemporary chimneys are generally of circular cross-section and are of metal fabrication. An important consideration in chimney design not discussed here is the wind loads imposed on the structure, which can be carried by the fabric of the chimney itself or by external bracing such as fins or wires. In addition to the steady-state loads on a chimney, vibrations can be induced by the regular shedding of vortices from the cylindrical surface (the same mechanism which causes telephone wires to hum). The spiral 'strakes' which are often attached to the upper part of a chimney are there to reduce this effect.

The force for moving the flue gas within the system can come from the buoyancy of the hot gas within the flue, from external fan power or from a combination of both. In addition to maintaining the correct flow rate of gas through the flue it is essential to maintain the temperature of the gas within the flue system above the water vapour dew point, or the acid dew point if a high sulphur content fuel is being burned. In terms of mechanisms, we are interested in the heat transfer and fluid flow performance of the flue, recognising that both these considerations must be integrated in the engineering design of the flue system.

# 11.2 Chimney Heat Transfer

## 11.2.1 Heat Transfer Mechanism

It is important to be able to quantify the heat transfer mechanism which operates in a chimney system. The rate of heat transfer affects the temperatures in the system, which in turn determine the safety margin in respect of possible corrosion problems, and also the pressure difference due to buoyancy forces, known as the chimney draught. The most common situation is where the temperature, composition and flow rate of the gas entering the flue are known, and the objective is to find the temperature of the gas as it is discharged from the flue.

The approach adopted here is to consider the flue in classical heat exchanger terms. This approach involves two stages: evaluating an overall thermal conductance (*U*-value) followed by an analysis of the performance of the flue, taking into account the flows of the system fluids (flue gas and ambient air).

## 11.2.2 *U*-value of a Chimney

Consider the cross-section sketch of a chimney wall shown in Fig. 11.1. There are three sequential stages in the steady-state heat transfer process:

1. Convective heat transfer from the hot flue gas to the inside surface of the chimney.
2. Heat transfer between the inside and outside surfaces of the chimney. If the chimney is of solid construction, the mechanism for this will be conduction. If this is not the case, for instance if there is an air gap present, then a combined mode of heat transfer (conduction/convection together with radiation) will be operating.
3. Convective heat transfer from the outside surface of the chimney to the ambient air. Under still air conditions this will be by natural convection, but in general the action of wind will induce forced convection from the chimney.

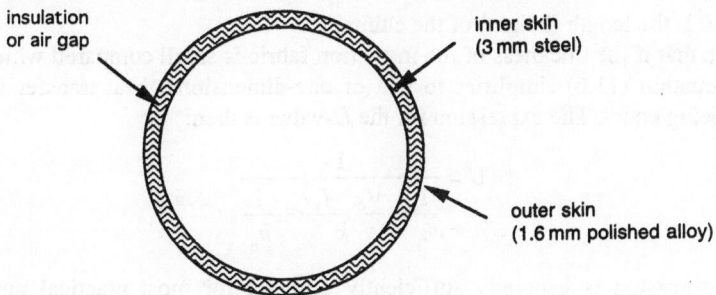

insulation or air gap

inner skin (3 mm steel)

outer skin (1.6 mm polished alloy)

**Figure 11.1** Chimney cross-section

These processes can be represented thus:

$$Q = h_i A_i (t_g - t_{si}) \qquad (11.1)$$

$$Q = h_f A_f (t_{si} - t_{so}) \qquad (11.2)$$

$$Q = h_o A_o (t_{so} - t_o) \qquad (11.3)$$

In equation (11.2) the heat transfer across the fabric has been expressed in terms of a fabric heat transfer coefficient, $h_f$, and a cross-sectional area, $A_f$, over which this operates. The contributory terms to $(h_f A_f)$ depend on the construction of the chimney, as will be shown later. The above equations can be summed giving:

$$Q\left(\frac{1}{h_i A_i} + \frac{1}{h_f A_f} + \frac{1}{h_o A_o}\right) = (t_g - t_o) \qquad (11.4)$$

The rate of heat transfer is conventionally represented in the form

$$Q = U_o A_o (t_g - t_o) \qquad (11.5)$$

where $U_o$ represents the overall heat transfer coefficient for the exchanger and $A_o$ represents the area which is associated with it. Any of the three areas $A_i$, $A_f$ or $A_o$ could be used for this purpose, but here the outside surface area of the chimney (the largest of the three) has been used. An expression for the overall $U$-value of the chimney is obtained by dividing (11.4) by (11.5):

$$U_o = \frac{1}{\dfrac{A_o}{h_i A_i} + \dfrac{A_o}{h_f A_f} + \dfrac{1}{h_o}} \qquad (11.6)$$

If the chimney is of circular cross-section and does not have an appreciable taper, the areas in the above expression are given by

$$A_o = 2\pi r_o L$$

$$A_i = 2\pi r_i L$$

$$A_f = \frac{2\pi (r_o - r_i)L}{\ln(r_o / r_i)}$$

where $L$ is the length (height) of the chimney.

Note that if the thickness of the insulation/fabric is small compared with the radius, then equation (11.6) simplifies to that of one-dimensional heat transfer with all the areas being equal. The expression for the $U$-value is then:

$$U = \frac{1}{\dfrac{1}{h_i} + \dfrac{(r_o - r_i)}{k} + \dfrac{1}{h_o}} \qquad (11.7)$$

This expression is generally sufficiently accurate for most practical purposes. For convenience, we can replace $(r_o - r_i)$ with $x$, the thickness of the fabric layer. The thermal resistance of this layer is then

$$\frac{x}{k} \qquad \mathrm{K\,m^{-2}\,W^{-1}}$$

and for a composite construction of $n$ layers, the total thermal resistance is found by adding up the individual layer resistances

$$\sum \frac{x_i}{k_i} \qquad \mathrm{K\,m^{-2}\,W^{-1}}$$

A common configuration for chimneys of less than 15 m height is to have an outer skin of aluminium alloy, with a low-emissivity surface, combined with a steel inner lining with an air gap of around 6 mm between the two metals. The heat transfer across an air gap is affected by the emissivity of the surfaces and the width of the gap. In general, the thermal resistance of an air gap increases with its width, but the value remains substantially constant at separations greater than 20 mm. The following values can be taken:

|  | Low emissivity surface | High emissivity |
|---|---|---|
| 6 mm | 0.18 | 0.1 |
| 20 mm + | 0.35 | 0.18 |

the units being $\mathrm{K\,m^2\,W^{-1}}$.

To evaluate (11.7), values are needed for the inside and outside surface coefficients of heat transfer, together with the thermal resistance of the chimney fabric and a means of estimating its effective heat transfer area.

### Inside Surface Coefficient

The flow of gas in the chimney may well be driven by buoyancy forces, but these are generated by the difference in density between the hot flue gas and the ambient air. Under these circumstances the contribution to the buoyancy force provided by the difference in temperature between the flue gas and the inside surface of the chimney will be very small, and heat transfer from the gas to the chimney inside surface will take place by forced convection.

The flue gas flow regime (whether the flow is laminar, transitional or turbulent) is indicated by the value of the Reynolds number

$$\mathrm{Re} = \frac{v d_i}{u}$$

where $u$ is the kinematic viscosity of the flue gas. For a circular cross-section the characteristic dimension $d$ is given by the diameter of the chimney. To get some perspective on this situation we can take a range of diameters between 0.25 m and 2 m, and consider the applicable range of flue gas velocity ($v$) to be between 6 m s$^{-1}$ and 20 m s$^{-1}$.

The kinematic viscosity of a gas varies with temperature, and, although values for individual gases are readily available [1], it is not possible to calculate accurately the viscosity of a mixture from the values and volume fractions of its constituents. The estimated kinematic viscosity of a flue gas resulting from the combustion of a natural gas with 20% excess air is compared with the values for air in Table 11.1. It can be seen that there is comparatively little difference between the two gases at typical flue temperatures of around 200°C.

Table 11.1 Kinematic viscosities for typical flue gas and air

| Temperature (°C) | Kinematic viscosity (m² s⁻¹ × 10⁶) | |
|---|---|---|
| | Flue gas | Air |
| 50 | 17 | 18 |
| 100 | 22 | 23 |
| 150 | 27 | 27 |
| 200 | 33 | 32 |
| 250 | 40 | 36 |
| 300 | 46 | 41 |
| 350 | 54 | 46 |

At a temperature of $t$°C the kinematic viscosity of air is approximated by

$$u = (0.1335 + 0.925 \times 10^{-3} \, t) \, 10^{-4} \qquad m^2 \, s^{-1} \qquad (11.8)$$

giving a value at 200°C of $0.000\,032 \, m^2 \, s^{-1}$.

The Reynolds number for a flue gas velocity of 9 m s⁻¹ in a chimney of 750 mm diameter is thus

$$Re = \frac{9 \times 0.75}{32 \times 10^{-6}}$$

$$= 210\,900$$

which is well into the turbulent regime.

The emissivity of the flue gas is low, hence the predominant mode of heat transfer is forced convection. A relationship for forced convection from a turbulent gas flow inside a cylindrical tube is [2]

$$Nu = 0.023 (Re)^{0.8} (Pr)^{0.4} \qquad (11.9)$$

where the Nusselt number, Nu, is given by

$$Nu = \frac{h_i \, d_i}{k}$$

and the Prandtl number, Pr, by

$$Pr = \frac{c_p \, \mu}{k}$$

Most gases have a value of Pr of about 0.74, and this value is substantially independent of temperature, hence equation (11.9) simplifies to

$$\text{Nu} = 0.02 \, (\text{Re})^{0.8} \qquad\qquad (11.10)$$

and the coefficient of heat transfer is obtained knowing the diameter of the tube, $d$, and the thermal conductivity of the gas, $k$.

Once again, a value for $k$ for the gas mixture could be estimated from a knowledge of the individual gas properties and the mixture composition. Here it is also an acceptable approximation to use values for air. The thermal conductivity of air can be obtained from

$$k_{air} = 0.02442 + 0.6992 \times 10^{-4} \, t \qquad \text{W m}^{-1} \text{K}^{-1}$$

where $t$ is the temperature in °C. At 200°C this gives a value of 0.0384 W m$^{-1}$ K$^{-1}$.

## Example 11.1

Estimate the internal coefficient of heat transfer in a 500 mm diameter chimney. The flue gas velocity is 10 m s$^{-1}$ and the gas temperature 250°C.

Start by working out the gas kinematic viscosity and thermal conductivity:

$$u = (0.1335 + 0.925 \times 10^{-3} \times 250) \times 10^{-4}$$
$$= 3.64 \times 10^{-5} \text{ m}^2 \text{ s}^{-1}$$
$$k = 0.02442 + 0.6992 \times 10^{-4} \times 250$$
$$= 0.0419 \text{ W m}^{-1} \text{K}^{-1}$$

$$\text{Re} = \frac{vd_i}{u}$$
$$= \frac{10 \times 0.5}{3.64 \times 10^{-5}}$$
$$= 1.37 \times 10^5$$

The Nusselt number, Nu, is

$$\text{Nu} = 0.02 \, (1.37 \times 10^5) \times 0.8$$
$$= 257$$

and the coefficient of heat transfer

$$h_i = \frac{\text{Nu}\,k}{d_i}$$
$$= \frac{257 \times 0.0419}{0.5}$$
$$= 21.6 \, W m^{-2} \, K^{-1}$$

An approximate value of $h_i$ can be obtained more quickly from Fig. 11.2, where values of $h_i \times d_i$ (internal film coefficient × chimney diameter) are plotted as a function of $v.d_i$ (flue gas velocity × chimney diameter) for flue gas temperatures ranging from 100 to 300°C.

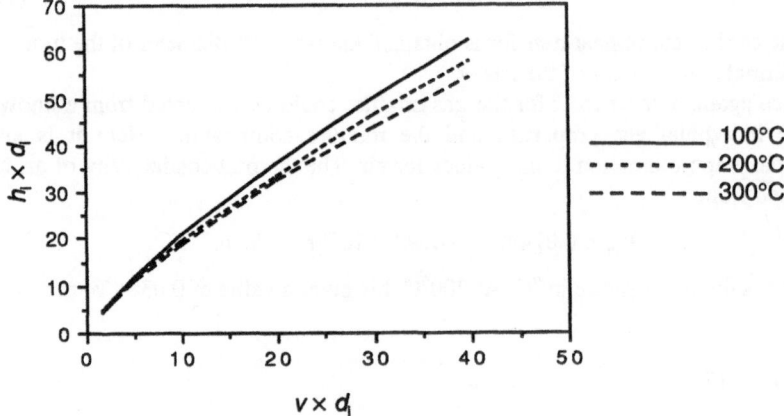

**Figure 11.2**   Inside surface film coefficient

For the above example $v \times d_i = 10 \times 0.5 = 5 \text{ m}^2 \text{ s}^{-1}$, giving a value of $h_i \times d_i$ from the graph at 250°C of 10.8 W m$^{-1}$ K$^{-1}$, resulting in an estimated value of

$$h_i = 10.8 \div 0.5 = 21.6 \text{ W m}^{-2} \text{ K}^{-1}$$

## Outside Surface Coefficient

The heat transfer processes taking place on the outside surface of the chimney are rather more complex than is the case for the inside surface coefficient. Radiation heat transfer does have a part to play – the surface of the chimney will exchange heat by radiation to the surrounding environment, and the convective heat loss will be affected by the prevailing wind speed.

In calm conditions, the convective heat transfer from the outside surface of the chimney will be by natural convection. As the wind velocity increases, forced convection will become the dominant mechanism. Bearing in mind that the main objective of quantifying the heat loss rate from a chimney is to identify cool surface temperatures, it is a reasonable approach to analysing the problem to focus attention on the 'worst case' scenario. In this context the situation is represented by forced convection dominating the heat transfer from the outside surface of the chimney.

A chimney exposed to the wind approximates to a cylinder with its axis at right-angles to the direction of the flow. The relevant heat transfer relationship, valid for Reynolds numbers between $10^3$ and $10^5$ is:

$$\text{Nu} = 0.26 \, (\text{Re})^{0.6} \, (\text{Pr})^{0.3}$$

Assuming a constant value for Prandtl number for air of 0.74, this expression simplifies to:

$$\text{Nu} = 0.24 \, (\text{Re})^{0.6} \tag{11.11}$$

In this case the Reynolds and Nusselt numbers refer to the outside diameter, $d_o$, of the chimney. The outside convective heat transfer coefficient is then given by:

$$h_{c,o} = \frac{\text{Nu}\,k}{d_o} \tag{11.12}$$

For the case of forced convection the temperature of the air has only a small effect on the value of $h_o$, and Fig. 11.3 shows the variation of $h_{c,o}$ with windspeed for a range of chimney outside diameters.

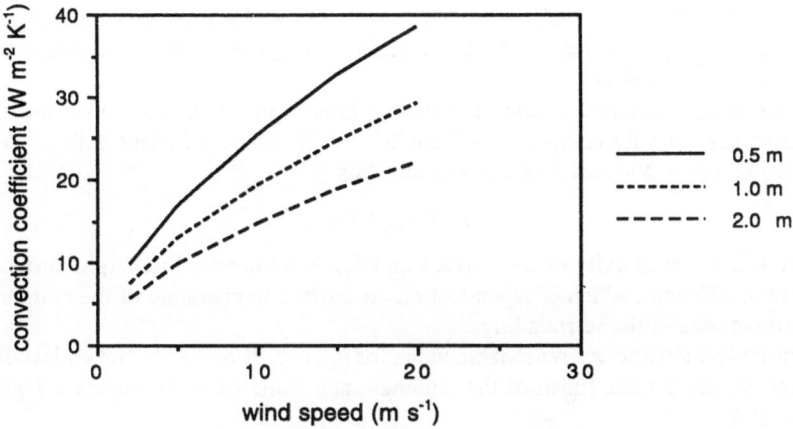

Figure 11.3   Outside surface film coefficient

## Example 11.2

Estimate the outside convective heat transfer coefficient for a 750 mm diameter chimney exposed to a wind of 10 m s$^{-1}$. Assume the air temperature to be 5°C.

At 5°C the kinematic viscosity of air from (11.8) is:

$$u = (0.1335 + 0.925 \times 10^{-3} \times 5)\,10^{-4}\ \text{m}^2\ \text{s}^{-1}$$

$$= 0.0000138\ \text{m}^2\ \text{s}^{-1}$$

The Reynolds number is then:

$$\text{Re} = \frac{10 \times 0.75}{13.8 \times 10^{-6}}$$

$$= 5.43 \times 10^5$$

$$\text{Nu} = 0.24\,(5.43 \times 10^5)^{0.6}$$

$$= 662.4\ \text{(from equation 11.11)}$$

The thermal conductivity of air at 5°C is

$$k = 0.02442 + (0.6992 \times 10^{-4} \times 5)$$

$$= 0.0248 \text{ W m}^{-1} \text{ K}^{-1}$$

giving an outside convective film coefficient of

$$h_{c,o} = \frac{662.4 \times 0.0248}{0.75}$$

$$= 21.9 \text{ W m}^{-2} \text{ K}^{-1}$$

The heat transfer from the outside surface of the chimney to the environment also contains a contribution from a radiative exchange to the surroundings. This is not an easy mechanism to quantify accurately, as radiative heat transfer is a non-linear process with respect to temperature and is affected by the nature and configuration of the participating surfaces.

However, in circumstances such as these a linear approximation can be made to the radiative heat transfer component [3] and an outside film coefficient defined in terms of the convective and radiative components thus

$$h_o = h_{c,o} + \varepsilon \, h_{r,o}$$

where $\varepsilon$ is the emissivity of the surface and $h_{r,o}$ is a linearised radiation surface heat transfer coefficient, which is dependent on the surface temperature of the chimney and the temperature of the surroundings.

For typical situations a reasonable value for $h_{r,o}$ is 5 W m$^{-2}$ K$^{-1}$. The emissivity will depend on the outside finish of the chimney, and some example values are given in Table 11.2.

**Table 11.2** Emissivities of some surfaces

| Material | Emissivity |
|---|---|
| Polished aluminium | 0.10 |
| Oxidised aluminium | 0.18 |
| Polished steel | 0.07 |
| Rolled sheet steel | 0.66 |
| Galvanised zinc | 0.21 |
| Brick | 0.93 |
| Glazed surface | 0.90 |
| Non-metallic paint | 0.90 |

If we want to take into account the effects of radiation in Example 11.2, for the case of a dull metal finish ($\varepsilon = 0.2$), we have:

$$h_o = 21.96 + 0.2 \times 5$$

$$= 22.96 \text{ W m}^{-2} \text{ K}^{-1}$$

If the chimney is insulated, the contribution of variation in $h_o$ to the overall heat transfer process is quite small, hence the approximations introduced above are justifiable.

## Overall $U$-value

Bringing the inside and outside film coefficients together with the thermal resistance of the chimney fabric into the evaluation of the overall $U$-value of a chimney is best illustrated by a simple example.

## Example 11.3

A flue has a height of 15 m, an external diameter of 750 mm and a construction consisting of an internal steel lining, 3 mm thick, of 50 mm mineral wool insulation (conductivity 0.04 W m$^{-1}$ K$^{-1}$), and an outer skin of 1.6 mm alloy with a polished outer surface. The internal and external heat transfer coefficients have been estimated as 20 and 23 W m$^{-2}$ K$^{-1}$ respectively.

From equation (11.7), neglecting the thermal resistance of the two metal layers:

$$U_o = \frac{1}{\dfrac{1}{20} + \dfrac{0.05}{0.04} + \dfrac{1}{23}}$$

so  $U_o = 0.746 \, W \, m^{-2} \, K^{-1}$

The relative magnitudes of the three terms in the denominator of the above expression show that by far the most significant term in the case of an insulated flue is the thermal resistance to conduction across the insulated layer, and thus in these circumstances a high level of accuracy in the evaluation of the film coefficients is unwarranted.

## 11.2.3 Temperature Distribution in a Chimney

In order to evaluate the overall heat transfer from the flue to the surroundings it is necessary to take into account the flow pattern of the two fluids concerned, i.e. the flue gases rising in the flue and the air flowing over the outside of the chimney. The latter can be considered as being at a constant temperature $t_o$, hence we can construct an energy balance about a short section of the chimney:

$$dQ = -U \, (t - t_o) \, dA \qquad (11.13)$$

$$dQ = W \, dt \qquad (11.14)$$

where $dA$ represents the surface area of the small section under consideration and $dt$ is the small temperature drop of the flue gases in this section. The term $W$ represents the thermal capacity rate of the flue gas, and is obtained by summing the product of the mass flow rate and specific heat for each of the species present, that is:

$$W = \Sigma(m\, c_p) \quad \text{kW K}^{-1}$$

Evaluation of this expression can be a bit time-consuming. However, an approximate value of 1.476 kJ m$^{-3}$ K$^{-1}$ can be assumed for the volumetric specific heat of the flue gas which, when multiplied by the volume flow rate of the gas (m$^3$ s$^{-1}$) gives a value for $W$. The volume flow rate of flue gas can be estimated from the gas velocity and the internal cross-sectional area of the chimney.

Equation (11.13) is a rate equation and (11.14) is an energy balance on the gas in the control volume; note that in these equations, heat lost from the gas is treated as a negative quantity. Eliminating d$Q$ from these equations gives

$$-U\,(t - t_o)\, dA = W\, dt$$

$$-\frac{U}{W} = \frac{dt}{(t_1 - t_o)}$$

which can be integrated

$$-\frac{U}{W}\int_0^A dA = \int_{t_1}^{t_2} \frac{dt}{(t - t_o)}$$

giving

$$-\frac{UA}{W} = \ln\frac{(t_2 - t_o)}{(t_1 - t_o)}$$

Hence

$$t_2 = t_o + (t_1 - t_o)e^{-(UA/W)} \tag{11.15}$$

This expression gives the flue gas temperature $t_2$ at the outlet of a section of the flue of area $A$, from a starting temperature of $t_1$. It is, of course, most appropriately applied to the entire flue but the expression can be solved for any number of flue sections, so giving a temperature distribution along the flow path.

As the thermal capacity rate of the flue gas is quite high, the ratio $UA/W$ is generally small, giving a low temperature drop in the flue gas as within the chimney. The main function of the chimney insulation is to keep the temperature of the gas high by limiting the temperature drop between the gas and the inner lining of the chimney.

In the steady state, the ratio of the temperature drop across the inner boundary layer to the overall temperature difference between the gas and the outside air is equal to the ratio between the thermal resistance of the inner boundary layer to the resistance of the entire chimney wall. Noting that thermal resistance is the reciprocal of conductance we can write

$$\frac{t_g - t_s}{t_g - t_o} = \frac{U_o}{h_i}$$

hence

$$t_s = t_g - \frac{U}{h_i}(t_g - t_o) \tag{11.16}$$

Maintaining internal surfaces well above the gas dew point is an important design consideration. Estimation of the surface temperatures in a flue is illustrated in the following example.

## Example 11.4

Estimate the flue gas temperature on leaving and the internal surface temperature at the flue exit for a 15 m high flue with a $U$-value of 3.5 W m$^{-2}$ K$^{-1}$ at an outside air temperature of $-3°$C. The internal diameter of the flue is 1.25 m and the external diameter 1.5 m. The flue gas velocity is 4 m s$^{-1}$ and the gas temperature on entering is 275°C.

The first task is to find the bulk flue gas temperature at the chimney exit from equation (11.15). The capacity rate of the gas must first be obtained from the gas velocity and the internal cross-sectional area:

$$W = 1.467 \times \pi \times 1.25^2 \times 4$$

$$= 28.98 \ kW \ K^{-1}$$

The outside surface area, $A$, of the chimney is

$$A = \pi \times 1.5 \times 15$$

$$= 70.69 \ m^2$$

so
$$UA = 0.247 \ kW \ K^{-1}$$

The temperature of the gas at the flue exit is:

$$t_2 = -3 + (275 + 3)e^{-(0.247/28.98)}$$

$$= 272.6°C$$

The next step is to estimate the internal film coefficient – the approximate method outlined in the previous section will be used.

The value of $(v \, d_i)$ is $4 \times 1.25 = 5$, hence $(h_i \, d_i)$ will be 10.7 (from Fig. 11.2). Hence:

$$h_i = 10.7/1.25$$

$$= 8.56 \ W \ m^{-2} \ K^{-1}$$

We can now solve (11.16) to get the internal surface temperature:

$$t_s = 272.6 - \frac{3.5}{8.56}(272.6 + 3)$$

$$= 159.9°C$$

A $U$-value of 3.5 W m$^{-2}$ K$^{-1}$ is, of course, a fairly modest level of thermal insulation but would be typical of a double skin flue with a low-emissivity air gap as insulation. The example does, however, illustrate the importance of the temperature gradient *across* the chimney as opposed to that *along* the flow path of the flue gas.

# 11.3 Pressure Loss

## 11.3.1 Chimney Draught

The chimney contains a column of hot gas which has a density considerably lower than that of the air surrounding it (Fig. 11.4). This difference in density produces an apparent 'suction' at the base of the flue which is known as the 'chimney draught'. In some instances (mainly smaller combustion equipment) this pressure difference is sufficient to overcome the resistance to flow through the boiler and the flue, and to provide the correct efflux velocity at the top of the flue. In cases where this natural draught is insufficient, fan assistance can be employed. In the majority of installations, the combustion air is supplied to the burner by an integral fan, which will have sufficient duty to propel the air through the boiler itself. In these circumstances all the flue draught is available for providing the energy required to move the flue gas through the chimney.

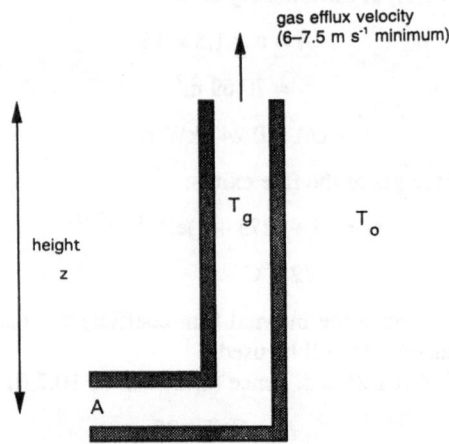

**Figure 11.4** Chimney buoyancy forces

Taking the point A in Fig. 11.4, the pressure difference due to the buoyancy of the column of hot gas in the chimney is simply equal to the relative weights of the columns of hot gas and surrounding air:

$$\Delta p = zg \, (\rho_g - \rho_a) \tag{11.17}$$

In this expression $\rho_g$ represents the mean density of the flue gas. As was shown above, the temperature drop as the gas passes through the chimney is quite small, so there would be only a small error incurred if the temperature of the gas at the flue inlet was used to evaluate the available draught.

The density of the air is primarily influenced by its temperature, but the density of the flue gas is dependent on both the temperature and composition of the gas, which will vary with different fuels and the air-to-fuel ratio.

The equation of state for an ideal gas

$$pV = \frac{m}{M} RT$$

can be written in terms of the gas density, $\rho$

$$p = \frac{\rho}{M} RT$$

or

$$\rho = \frac{pM}{RT}$$

At standard atmospheric pressure of 1.013 25 bar and noting that the universal gas constant, $R$, is 8.314 kJ kmole$^{-1}$ K$^{-1}$:

$$\rho = 12.1873 \frac{M}{T} \quad \text{kg m}^{-3} \tag{11.18}$$

This expression can be substituted for the two density terms in (11.17), noting that the average molecular weight for air is 28.84 and incorporating $g$ in the numerical constant:

$$\frac{\Delta p}{z} = 118.9 \left( \frac{M_g}{T_g} - \frac{28.84}{T_a} \right) \quad \text{Pa m}^{-1}$$

A further simplification can be made if we consider the molecular weight of the flue gas, $M_g$. This will obviously depend on the chemical constitution of the fuel and on the air-to-fuel ratio. Some typical values for $M_g$ range from 27.86 (natural gas burned at 20% excess air) to 30.0 (bituminous coal burned at 20% excess air). Within this range it is reasonable to take an approximate value equal to the mean molecular weight of air, i.e. 28.84, leading to the simplified expression for the chimney draught:

$$\frac{\Delta p}{z} = 3429 \left( \frac{1}{T_g} - \frac{1}{T_a} \right) \tag{11.19}$$

As an example, if the ambient temperature is $-5°C$ (268 K) and the mean flue gas temperature is $250°C$ (523 K), then the draught, given by equation (11.18), is $-6.24$ Pa m$^{-1}$.

## 11.3.2 Flow Resistance

In a natural-draught appliance, the buoyancy force produced in the flue has to move the air through the boiler and the flue system. Where the burner incorporates a forced air supply via a fan, it is most likely that the designer will have to ensure that an adequate pressure difference is available to overcome the resistance of the flue system itself. The basis for quantifying the pressure (or energy) losses when a fluid flows through a system containing resistances due to bends, fittings and the friction of the pipe itself is

$$\Delta p = K \, (v^2/2g) \qquad (11.20)$$

where $K$ is the pressure loss coefficient and the bracketed term is the velocity pressure of the fluid. For the case of the friction in the conduit of the chimney itself $K$ is represented as

$$K = f(L/D)$$

where $f$ is the friction factor which depends on both the roughness of the surface and the Reynolds number of the flow. For a fuller treatment of the relationships for pressure loss in pipes and ducts, the reader is referred to any standard text on fluid mechanics or the CIBSE [4] and ASHRAE [5] handbooks.

If an estimate of the flue gas velocity is available, an initial estimate of the pressure loss in the flue system can be made by substituting a value for $K$ of 5.0 in equation (11.20). At a later stage, when the design of the flue system is more established, the pressure drop for the system can be obtained from summing up the individual contributions to the pressure loss multiplied by their respective velocity pressures:

$$\Delta p = (K_1 + \Sigma K_f + K_D) \, vp_1 + K_2 \, vp_2 \qquad (11.21)$$

Here $K_1$ is the loss coefficient at the inlet to the flue duct. The second term represents the summation of all the individual losses from fittings such as bends, dampers or connectors. For values of these, the reader is referred again to the handbooks; approximate example values are 0.8 for a smooth round 90° bend and 1.25 for a mitred 90° bend. For the duct loss $K_D$ an estimate can made from:

$$K_D = L/(30 \, d_i)$$

With regard to the exit loss $K_2$ is equal to 1.0 and is associated with the velocity pressure in the chimney duct $vp_1$ if the chimney discharges as a free jet into the atmosphere, but there will be an additional contributing factor (based on the velocity pressure $vp_2$ in the smaller outlet diameter) if a tapering conical cap is used to increase the efflux velocity of the jet. A typical value of $K_2$ for a gradual contraction is 0.2.

## Example 11.5

For the chimney described in Example 11.4, find the chimney draught and the pressure drop in the flue if a cap is used to accelerate the efflux velocity to 8 m s$^{-1}$.

In Example 11.4, the gas temperature on entering was 275°C and the gas temperature on leaving was 272.6°C. The mean temperature is thus 273.8°C (547 K). The outside air temperature is −3°C (270 K), hence the draught, from equation (11.19) is:

$$\frac{\Delta p}{z} = 3429 \left( \frac{1}{547} - \frac{1}{270} \right) \quad \text{Pa m}^{-1}$$

$$\Delta p = -96.5 \, Pa$$

To evaluate the pressure drop it is first necessary to estimate the density of the flue gas. This is done via equation (11.18):

$$\rho = 12.1873 \frac{M}{T}$$

Assuming a molecular weight of 29, this gives:

$$\rho = 12.1873 \frac{29}{547}$$

$$= 0.646 \text{ kg m}^{-3}$$

The velocity pressure in the flue duct is then:

$$vp_1 = 0.646 \frac{4^2}{2}$$

$$= 5.2 \text{ Pa}$$

Assuming a friction factor of 0.033, the pressure loss coefficient for the duct, $K_D$, is:

$$K_D = 0.333 \frac{15}{1.25}$$
$$= 0.396 (0.4)$$

If the flow is accelerated to 8 m s$^{-1}$ at the outlet, the velocity pressure is:

$$vp_2 = 5.2 \times 8^2/4^2$$

$$= 20.8 \text{ Pa}$$

Assuming a $K$-factor of 0.2 for the reducer, and 1.0 for the discharge of the free jet, the total pressure drop is given by:

$$p = (0.4 \times 5.2) + (1.2 \times 20.8)$$

$$= 27 \, Pa$$

Integrating the heat transfer and pressure loss characteristics into a design procedure is, of necessity, an iterative process, as are so many engineering design tasks. It is beyond the scope of this book to stray into this territory, but it is suggested that a convenient starting point for the iteration is to assume a trial value for the flue gas velocity. This leads to a preliminary estimate for the internal diameter of the chimney and a first set of values for the hydraulic and heat transfer performance of the flue system. However, it is perhaps a little premature to embark on this without also giving some consideration to the location and height of the chimney.

# 11.4  Chimney Location

The principal factor affecting the height and location of a chimney is the safe and effective disposal of the products of combustion. This in turn is not simply concerned with ground level concentrations of pollutants; it is important to ensure, for example, that combustion products are not re-ingested into the intakes of combustion systems or the fresh air inlets of HVAC systems. Other considerations are entry of debris, fire risks, ingress of rain and wind effects on the exit flow.

The flow of air around the building in which the plant is located plays a definitive role in the design of a flue system. As the wind flows over the ground, a velocity boundary layer is formed owing to the drag exerted by the ground on the air flow. The layers of air very close to the ground will have a very low windspeed, and this speed will increase up to a height where the flow is unobstructed. This depth over which the windspeed is changing is called the boundary layer, and for flow over developed terrain the boundary layer can be very thick, up to 500 m in some urban locations. The consequence of this is that buildings are immersed in the boundary layer (Fig. 11.5) and so the pressure generated when the wind hits the windward face of the building will increase with height.

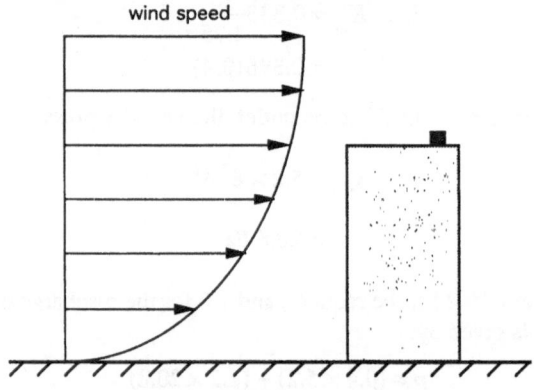

wind speed

**Figure 11.5**  Wind velocity gradient

As a result of this, wind flows down the front face of a building establishing a recirculating eddy on the windward face of the building. A separation bubble is formed where the flow separates from the top edge of the roof and there is a large recirculating wake formed in the lee of the building (Fig. 11.6). This basic flow pattern heavily influences the location of the flue outlet.

Clearly a low outlet (position A in Fig. 11.6) is unsatisfactory as the products of combustion will be trapped in the roof separation bubble. A higher location such as B should minimise problems of re-ingestion into other systems but there would still be a possibility of poor outdoor air quality in the leeward wake of the building. A high outlet such as C will clearly avoid all these problems, discharging the combustion products into the streamlines which are substantially unaffected by the building. A high discharge velocity from the open end of the flue will clearly aid dispersion, but this can only be achieved if fan assistance is available.

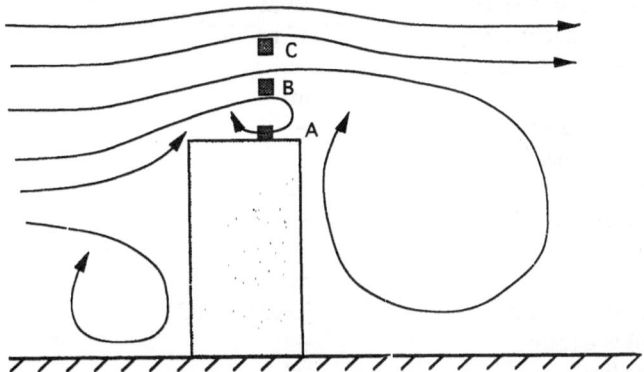

**Figure 11.6**   Air flow patterns around a building

High discharge points, while being the best solution from a dispersion point of view, will have an improved draught but will also need more insulation for their greater exposed surface area and will generate higher wind loads. Many designers and architects are not enthusiastic about the visual attributes of relatively tall chimneys. There are also many potential complicating factors. If the roof is extensive in the horizontal direction, the separation bubble may reattach itself to the roof surface. The immediate environment of the building (other buildings, trees or even nearby hills) may also influence the local aerodynamics. In such difficult cases, it may be necessary to resort to scale model testing of the design proposals in a boundary layer wind tunnel.

The nature of the fuel also plays an important part in locating the flue outlet. If the fuel is sulphur-free, such as natural gas, it may be possible to dilute the combustion products with air and discharge them at a low level close to pedestrian areas. If there is a significant sulphur emission in the flue gas, then statutory requirements such as the UK Clean Air Act will dictate the height of the discharge.

Because of the complicated nature of this problem, it is not possible to give any simple rules of thumb. Design procedures are described, for example, in the CIBSE Guide [4] together with some worked examples. In the context of the building flow patterns shown in Fig. 11.6, the height of the recirculation eddy above the roof is very roughly equal to twice the building height, although it should be noted that this figure is also influenced by the horizontal dimensions of the building.

## 11.5  List of Symbols

| | | |
|---|---|---|
| $A$ | area | $m^2$ |
| $c_p$ | specific heat at constant pressure | $kJ\ kg^{-1}\ K^{-1}$ |

| | | |
|---|---|---|
| $d$ | diameter | m |
| $f$ | friction factor | – |
| $g$ | gravitational constant | $m^2\,s^{-1}$ |
| $h$ | heat transfer coefficient | $W\,m^{-2}\,K^{-1}$ |
| $k$ | thermal conductivity | $W\,m^{-1}\,K^{-1}$ |
| $K$ | pressure loss coefficient | – |
| $L$ | length | m |
| $M$ | molecular weight | – |
| Nu | Nusselt number | – |
| $p$ | pressure | Pa |
| Pr | Prandtl number | – |
| $Q$ | heat flux | W |
| $r$ | radius | m |
| $R$ | universal gas constant | $kJ\,kmol^{-1}\,K^{-1}$ |
| Re | Reynolds number | – |
| $t$ | temperature | °C |
| $T$ | temperature | K |
| $u$ | kinematic viscosity | $m^2\,s^{-1}$ |
| $U$ | overall thermal transmittance | $W\,m^{-2}\,K^{-1}$ |
| $v$ | velocity | $m\,s^{-1}$ |
| $vp$ | velocity pressure | Pa |
| $V$ | volume | $m^3$ |
| $x$ | thickness | m |
| $z$ | height | m |
| | | |
| $\varepsilon$ | emissivity | – |
| $\mu$ | dynamic viscosity | Pa s |
| $\rho$ | density | $kg\,m^{-3}$ |

*Subscripts*

| | |
|---|---|
| f | fabric |
| g | gas |
| i | inside, inlet |
| o | outside |

# 11.6 References

1.  Spiers HM (1962). Technical Data on Fuel. British National Committee, World Power Conference, London
2.  McAdams WH (1954). Heat Transmission, 3rd edn. McGraw-Hill, New York

3.  CIBSE Guide (1986). Chartered Institution of Building Services Engineers, London, Section A3-8

4.  CIBSE Guide (1986). Chartered Institution of Building Services Engineers, London, Section B13

5.  ASHRAE Handbook Fundamentals (1989). American Society of Heating, Refrigerating and Air Conditioning Engineers, Atlanta, Georgia

# Index